A CHECK LIST OF THE FERNS FERN ALLIES, AND GYMNOSPERMS OF THE N.E. KEPALA BURUNG (VOGELKOP), IRIAN JAYA, INDONESIA

A REPORT TO THE

JOHN D. and CATHERINE T. MACARTHUR FOUNDATION

R.J. Johns, P.J. Edwards and D. Kowalska
Royal Botanic Gardens, Kew.

Frontspiece: Drawn by **Ann Farrer**

Papuacedrus papuanus (F.Muell.) H.L.Li
Curtis's Botanical Magazine. Plate 266.

A Checklist of the Fern Allies, Ferns and Gymnosperms of the N. E. Kepala Burung (Vogelkop), Irian Jaya, Indonesia by R. J. Johns, P. J. Edwards & D. Kowalska

ISBN 978-9980-945-92-1

University of Papua New Guinea Press and Bookshop
PO Box 413
University PO, NCD
Papua New Guinea

INTRODUCTION

This check list includes three unpublished lists of the collections of the fern allies, ferns and gymnosperms from the N.E. Kepala Burung (Vogelkop), Irian Jaya, Indonesia. The lists will be submitted for publication when the collections have all been examined in detail and the entries of collections in The Natural History Museum, Leiden and Bogor verified for specimens which are not duplicated in Kew. It will also be necessary to check in more detail the distribution and ecology of the species from all the listed herbaria. Where adequate material exists new species will be described. Material of all specimens is duplicated in Bogor and Manokwari. We wish to thank various botanists for their useful comments on the collections.

Collections listed in both the Sands and Johns series were made with collectors from Herbarium Bogoriense and the University of Cenderawasih, Manokwari Herbarium. The associated collectors are listed on the labels in the herbarium but, to conserve space, are not listed in the check list. All Sands and Johns collections are deposited in Bogor (BO), Manokwari (MAN) and Kew (K). Where suffificent material was collected duplicates will be sent to Leiden (L), Bishop (BISH) and Lae (LAE).

The only previous study of this group of plants from the Vogelkop was published by Gibbs in 1917. A comparison is made of the numbers numbers of species recorded in 1917 and that in the present report for each group.

	Gibbs (1917)	This Report	(estimated)
Fern Allies	14	27	(57)
Ferns	82	275	(375)
Gymnosperms	8	27	(4)

It should be noted that there are over 250 unidentified collections of ferns and over 60 unidentified collections of ferns allies which have yet to be studied. These occur in several difficult groups such as the *Hymenophyllaceae*, *Asplenium* and *Selaginella.* There is probably over 100 additional tax of ferns and at least 30 new taxa of fern allies to add to these lists from existing collections. Further collections, particularly from the montane zone, will add significantly to the number of species particularly in the families Grammitidaceae and Hymenophyllaceae. Once the flora is well known we should have in excess of 750 species of ferns from this area.

We wish to thank Jill Marsden for her work on the data base. Mark Coode has been of continuous assistance and encouragement.

R.J. Johns
Royal Botanic Gardens, Kew

CONTENTS

FERN ALLIES

R.J. Johns

PSILOTACEAE

Johns, R.J. (1981). The ferns and fern allies of Papua New Guinea. Part 9: The Psilotaceae. *P.N.G. Univ. Tech. Res. Rep. R* 48-81: 1.1-1.7.

PSILOTUM Sw.

Psilotum complanatum Swartz *Syn. Fil.* 117 (1806). Johns (1981: 1.4).
Psilotum flaccidum Wall.
Dist.: Widespread throughout the tropics. Papuasia to Hawaii and Australia.
Ecol.: Epiphytic in lowland rainforest and montane forest from sea level to 2,200 m. In Koebre, epiphyte in forest. 2250 m.
Specimens: ARFAK MTS.: Hatam - *Beccari s.n.* (K). KOEBRE MTS.: *Gibbs* 5633 (K). WITHOUT LOCALITY: *Gibbs* 6273 (K); *van Royen & Sleumer* 7133; BW 7129.
Notes:

Psilotum nudum (L.) Palisot de Beauvois *Prodr. Mousses & Lycopodes* 112 (1805). Johns (1981: 1.3).
Lycopodium nudum L. *Sp. Pl.* 1100 (1753).
Psilotum triquenatum Swartz.
Dist.: Widespread in tropical and temperate areas. Throughout Papuasia.
Ecol.: Terrestrial or epiphytic in mangroves, lowland rainforest and montane forest from sea level to 2,400 m. Locally common. Collected as an epiphyte on sago, and in crevices on limestone cliffs. Often as an epiphyte on tree ferns.
Specimens: BW 8569 (LAE). WITHOUT LOCALITY: BW 11340 (LAE); *van Royen & Sleumer* 7962.
Notes: This species is widely cultivated in Japan where many horticultural varieties are propogated.

Psilotum **spp.**
Specimens: WITHOUT LOCALITY: *van Royen & Sleumer* 8068; *van Royen & Sleumer* 8163; BW 10298.

EQUISETACEAE

EQUISETUM

Hauke, R.L. (1961-1962). A taxonomic reorganisation of *Equisetum*, subgenus *Hippochaete*, I - V. *Amer. Fern J.* **51**: 131 - 137; **52**: 29 - 35, 57 - 63, 123 - 130.

Hanke, R.L. (1963). A taxonomic monograph of the genus *Equisetum* subgenus *Hippochaete*. Bieh. *Nova Hedwigia* **8**: 1 - 123.

Page, C.N. (1972). An assessment of inter-specific relationships in *Equisetum*. *New Phytol.* **7**: 355 - 369. f. 1-2, pl. 1-4.

Equisetum ramosissimum Desf. *Fl. Atlant.* 2 (1799) 398. K.Sch. & Laut. (1900: 148); Laut. (1930: 470); Hanke (1963: 33).

Equisetum debile Roxb. *Vauch. Mon. Presles* (1822) 387. t.26.

Equisetum debile *var. minus* Milde *Ann. Mus. Bot. Lugd.-Bat.* 1 (1864) 60. **Type**: ex Java, Milgeenia, Timor. K.Sch. & Laut. (1900: 148); Laut. (1929: 2).

subsp. ***debile*** Haenke *Am. Fern. J.* **52**: 33 (1962). Muell. (1877: 102); Muell. (1889: 39); Copeland (1929: 272); Milde (1803: 59); Lauterbach (1930: 436); Lam (1924: 6); Baker (1887: 5); deVol (1964:); Haenke (1963: 61). **Type**:

Hippochaete debilis Seem. *Fl. Vit.* (1873) 424.

Dist.: Sri Lanka and Himalayas through the Malesia to Fiji. In Papuasia NW NE SE New Guinea.
Specimens: ARFAK MTS.: Mupi Dessa - *Sands* 6722. ANGGI LAKES: BW 276; BW 13536. **IRAY**: *Widjaja* 4340; *Johns* 7954. WITHOUT LOCALITY: *van Royen & Sleumer* 3999; *van Royen & Sleumer* 4996; *van Royen & Sleumer* 6731; *van Royen & Sleumer* 7020; BW 11496; BW 10360; *Johns* 7852.

LYCOPODIACEAE

Huperziaceae
Urostachyaceae Rothm. nom. illeg.

Herter, W. (1909). Beitrage zur Kenntnis der Gattung *Lycopodium*. Studien uber die Untergattung *Urostachys*. *Bot. Jahrb. Suppl.* **98**: 1-56.

Herter, W. (1916). Beitrage zur Flora von Papuasiens V. Die Lycopodin von Deutch Neu Guinea. *Bot. Jahrb.* **54**: 226-238.

Herter, W. (1949). An index to genera and species, with indication of synonomy, some new combinations: genera *Lycopodium* and *Urostachys*. *Ind. Lycop.* (1949) i - iv, 1 - 120.

Nessel, H. (1934). Neue Lycopodien, die von allen schon bekannten Arten durch ihren Habitus ganz besonders und auffallend sind. *Fedde Rep.* **36**: 177-193.

Nessel, W. (1935). Beitrage zur Kenntnis der Battung Lycopodium. Fedde Rep.**39**: 61-71.

Nessel, H. (1939). *Die Barlappgewachse (Lycopodiaceae) Eine beschreibende Zusammenstellung mit besonderer Beruckiichtigung ihrer Varietaten und Formen*. Jena (Gustav Fischer). i - viii, 1 - 404, f. 1 - 87.

Ollgaard, B. (1989). Index of Lycopodiaceae. *Biol. Skrifter* **34:** 1 - 135.

The following collections of members of the Lycopodiaceae were generally given field identifications as *Lycopodium* but include material from all three genera.

Specimens: ARFAK PLAINS: Mountains to south - *Sands* 6447. ARFAK MTS.: Mokwam Trail - *Johns* 7737; *Johns* 7796. ANGGI LAKES: *Johns* 7969; *Johns* 7973. Iray - *Johns* 7981; *Johns* 7983; *Johns* 8080. KOEBRE - *Johns* 8090. SUNGAI ACEMO: *Sands* 6154. WITHOUT LOCALITY: *van Royen & Sleumer* 7042; *van Royen & Sleumer* 7043; *van Royen & Sleumer* 7114; *van Royen & Sleumer* 7206.

HUPERZIA Bernh.

Schrader *J. Bot.* **2** (1800 [1801]) 126

Lectotype: *Huperzia selago* (L.) C. Martius & Shrank. *Hort. reg. monac.* 3 (1829) 3. Lycopodium selago L. (1753). Selected by Rothmales (1944) 59 [Pichi Sermolli (1971) 142 - 143].

Huperzia coralium (Spring in Miquel) Holub Brause (1920) 222; Ridley (1916) 264; Warburg (1900) 99.

Lycopodium coralium Spring in Miquel Pl. Jungh. (1854: 273) **Type**: Junghuhn 291, Java. (Holotype: LG [Badre 1983: 4]; Isotype: K, L, P).

Urostachys coralium (Spring) Nessel (1939). (Herter 1949: 57)

Dist.: Malaya, Java, SW New Guinea, New Ireland, Samoa.

Ecol.:

Specimens: GEELVINK: Mandamui - *Beccari s.n.* (FL). WITHOUT LOCALITY: BW 6385; BW 9748.

Notes: Herter (1949: 13) lists as a synonym of *Urostachys phlegmaria*.

Ollgaard (1989) group H12a: group/subgroup *Huperzia phlegmaria*.

Huperzia nummularifolium (Bl.) Chambers, Jermy and Crabbe. Herter (1909) 37; Schltr. in Brause (1912) 4; K.Sch. & Laut. (1905) 49; Wagner & Grether (1948) 63; Warburg (1900) 98.

Lycopodium nummularifolium Blume *Enum. Pl. Jav.* 2 (1828) 263. **Type**: Blume s.n. Sallak, Java. Sheet numbers 908.326-17 and 908.342-266 (L).

Phlegmariurus nummularifolius (Bl.) Ching

Urostachys nummularifolius (Bl.) Herter (1923, 1949)

Dist.: Malesia, Penang, Java, Papuasia to Vanuatu. In Papuasia: NE New Guinea. Admiralty Islands.

Specimens: WITHOUT LOCALITY: BW 9749; *van Royen & Sleumer* 3899; *van Royen & Sleumer* 7040.

Notes: Ollgaard (1989) group H12i: the group of *Huperzia phlegmaria*, subgroup of *Phlegmaria nummularifolia*.

Huperzia phlegmaria (L.) Rothm. or P. Beauv. Anon. (1899) 122; Beccari (1880) 400; Beccari (1886) 53; Fosberg (1940) 110; Hieron in K.Sch. (1898) 87; Kuhn (1889) 16; Lauterbach (1929) 559; Lauterbach (1930) 432; Mueller (1876) 74; Warburg (1892) 12; Warburg (1900) 99.

Lycopodium phlegmaria L. *Sp. Pl.* (1753) 1101. **Type**: Uncertain. Malabria, Zeylona. Several collateral references are cited. Dill. Hist. musc. t.61, f.5 A, B, C. (1741), is eligible as a lectotype.

Lepidotis phlegmaria (L.) P. Beauv.

Phlegmariurus phlegmaria (L.) Sen & Sen

Urostachys phlegmaria (L.) Herter

Dist.: E Africa, Asia, Malesia to Polynesia (Type:ahiti and Hawaii?), Queensland and New Zealand. In Papuasia: W NE SE New Guinea (Hellwig 349: Finisterres). New Britain, Solomon Islands.

Specimens: WITHOUT LOCALITY: *van Royen & Sleumer* 7038; *van Royen & Sleumer* 7041; *van Royen & Sleumer* 7089; *van Royen & Sleumer* 7605; *van Royen & Sleumer* 7735; *van Royen & Sleumer* 7802; *van Royen & Sleumer* 5009; *van Royen & Sleumer* 8220; BW 5581.

Notes: Ollgaard (1989) group H12a: the group and subgroup of *Huperzia phlegmaria*.

Huperzia pinifolium Trevisan Gibbs (1917: 77); Gepp. in Gibbs (1917) 77; Mueller (1883) 20; Schltr. in Brause (1912) 4; K.Sch. & Laut. (1905) 49; Warburg (1900) 99.

Lycopodium pinifolium Blume (1828) non Kaulf. (1824). *Enum. Pl. Java* **2** (1828) 264. **Type**: Blume s.n. Javae sylvis arborum (L 908.343-178).

Lycopodium piscium (Herter) Tag. & Iwatsuki

Urostachys pinifolius (Blume) Herter

Urostachys piscium Herter

Dist.: Java, Borneo, Celebes to New Guinea. In Papuasia: W NE SE New Guinea.

Ecol.: Epiphyte in *Araucaria* forest near Lake. 2100 m.
Specimens: ANGGI LAKES: *Gibbs* 5936 (BM).
Notes: Ollgaard (1989) group H8: the group of *Huperzia hippuris.*

Huperzia serrata (Thunb. ex Murray) Trevisan Gepp in Gibbs (1917) 77; Herter (1909) 32; K.Sch. & Laut. (1900) 150; Warburg (1892) 12; Warburg (1900) 98.
Lycopodium serratum Thunb. *Fl. Jap.* (1784) 341. t.38. **Type**: Thunberg s.n. Japan (Holotype: UPS-Type: Thunb. 25333B).
Dist.: S.E. Asia, throughout Malesia to Polynesia (Hawaii) and Mascarene Islands. Mexico (Bourbon). In Papuasia: W NE New Guinea.
Ecol.: Epiphyte in forests. 2100 m.
Specimens: ANGGI LAKES: *Gibbs* 5725 (BM).
Notes: (Hellwig 317a - Finisterres).
Ollgaard (1989) group H1: the group of *Huperzia selago.*

Huperzia subfalciforme (Alderw.) Holub Dakkus (1930) 186; Lam (1945) 37, 52; Nessel (1939).
Lycopodium subfalciforme Alderw. *Bull. Jard. Bot. Btzg.* II 28 (1918) 44. **Type**: Branderhorst 24. Cult. in Bogor ex New Guinea. (Holotype: BO, BM photo. fragment).
Urostachys subfalciforme (Alderw.) Nessel
Dist.: N.W. New Guinea.
Ecol.: Epiphyte near Pionierbivak and Prauwenbivak.
Specimens: WITHOUT LOCALITY: *van Royen & Sleumer* 5018.
Notes: Ollgaard (1989) group H12f: the group of *Huperzia phlegmaria* and subgroup of Huperzia phlegmarioides.

LYCOPODIELLA

Lycopodiella caroliniana (L.) Pichi Serm. *Webbia* **23**: 165 (1968). Ollgaard (1989: 25).
Lycopodium carolinianum L. Sp. Pl. 1104 (1753). **Type**: Dill. Hist. Musc. t.62, f.6 (1741). (OXF).
Pseudolycopodiella caroliniana (L.) Holub
Lepidotis caroliniana (L.) P.Beauv.
Specimens: ARFAK MTS.: *Kostermans* 2118.

Lycopodiella cernua (L.) Pichi Serm. *Webbia* **23**: 166 (1968). Ollgaard (1989: 25).
Lycopodium cernuum L. Sp. Pl. 1103 (1753). Gibbs (1917: 77). **Type**:
Dist.: Tropics and subtropics. In Irian Jaya: Arfak and Mt Jaya.
Ecol.: Common on open marsh and creeping in forest and on steep slopes.
Specimens: ARFAK MTS.: *Kostermans* 2435, Putat - *Beccari s.n.* (FL.). ANGGI LAKES: *Gibbs* 5658; *Gibbs* 5925 (BM); *Gibbs* 5995 (BM); BW 2019. WITHOUT LOCALITY: BW 9514; *van Royen & Sleumer* 7218; *van Royen & Sleumer* 7932.

Lycopodiella ericinum (Ces.) Pichi Serm.
Lycopodium ericinum Cesati Ollgaard (1989: 44).
Specimens: ARFAK MTS.: Hatam - *Beccari s.n.* (FL 12849).

LYCOPODIUM L.

L. *Sp. Pl.* (1753) 1100. et. *Gen. Pl.* ed.5 (1754).

Lycopodium casuarinoides Spring, *Bull. Acad. Roy. Soc. Belg.* **8**: 521 (1841). Gibbs (1917: 77); Herter (1949: 9); Posth. in Steen. (1934: 170); Ridley (1916: 265); Ollgaard (1989: 36). **Type**: *Cuming* 2346. Philippines. [Mt Ophir, Malacca]. (**Holotype** BM; K. L. P. UPS).

Lycopodiastrum casuarinoides (Spring) Dixit

Lepidotis casuarinoides (Spring) Rothm.

Dist.: Tropical Asia, India to S China, Malay Peninsula, Sumatra, Borneo, Philippines, Sunda Islands. W New Guinea.

Ecol.: Scrambling along edge of forest, 2100 m.

Specimens: ANGGI LAKES: *Gibbs* 5944 (BM). WITHOUT LOCALITY: *van Royen & Sleumer* 7153.

Notes: Ollgaard (1989) group L5: the group of *Lycopodium* Sect. *Lycopodiastrum*

Lycopodium clavatum L. **Sp. Pl.** (1753) 1101. **Type**: E. Cowley 29D. Upper Freshwater Creek, Queensland. Australia. (Holotype: BRI 5061; Isotype: K). Baker (1890) 110; Herter (1949) 11; Mueller (1889) 39; Nessel (1939) 287; Warburg (1900) 99.

Dist.: Cosmopolitan.

Lycopodium clavatum L. var. ***wallichianum*** Spring Gibbs (1917: 77); Nessel (1939: 296).

Dist.: Cosmopolitan in temperate areas.

Ecol.: Along open gravel bank and in open marsh. 2100 m.

Specimens: ARFAK MTS.: *Beccari s.n.* (FL.). ANGGI LAKES: *Gibbs* 5556 (BM); *Gibbs* 5911 (BM); BW 2049.

Notes: Several varieties have been described, most have been recorded from New Guinea. This includes var. *divaricatum* Wall. and var. *tamariscispica* Alderw.

Lycopodium complanatum L. *Sp. Pl.* (1753) 1104. Beccari (1886: 54); Gibbs (1917: 77); Lam (1924: 9); Nessel (1939: 327); Posth. in Steen. (1934: 170); Wilce (1965: 152, 220. t.39). **Type**:

Dist.: Pantropic and N Temperate area. In Malaysia, Malay Peninsula, Sumatra, Java, Philippines, Moluccas, NW New Guinea to Polynesia.

Ecol.: In open marsh on bank near lake. 2100 m.

Specimens: ARFAK MTS.: Hatam - *Beccari s.n.* (FL.); *Kostermans* 2190. ANGGI LAKES: *Gibbs* 5912 (BM); BW 10317.

Notes: *Lycopodium complanatum* var. *angustiramosum* (Alderw.) Herter, *Ind. Lycop.* 4. (1949). [*Lycopodium angustiramosum* Alderw. *Bull. Jard. Bot. Btzg.* II **24** (1917) 6] was described from the summit-area of Doormantop - Lam (1945: 112); Nessel (1939: 332. f.76).

Several varieties have been described in this species.

Lycopodium plegnum ?

Specimens: ARFAK MTS.: Putat - *Beccari s.n.* (FL.).

Lycopodium serratum Thunb. *Fl. Jap.* (1784) 341. t.38. Gibbs (1917: 77); Herter (1909: 32); K.Sch. & Laut. (1900: 150); Warburg (1892: 12); Warburg (1900: 98). **Type**:

Dist.: Mexico, Bourbon, S E Asia, throughout Malesia to Polynesia (Hawaii). In New Guinea: W and N.E.

Ecol.: Terrestrial in humus in forest. 2100 m.

Specimens: ANGGI LAKES: *Gibbs* 5724 (BM). WITHOUT LOCALITY: *van Royen & Sleumer* 3898; *van Royen & Sleumer* 7244; *van Royen & Sleumer* 7799; *van Royen & Sleumer* 8147; *van Royen & Sleumer* 8204.

Lycopodium volubile Forst. f. *Flor. Ins. Aust. od.* (1786) 86. Baker (1890: 110); Cesati (1877: 28); Gibbs (1917: 78). Mueller (1889: 39); Nessel (1939: 83. f.1). **Type**: Omitted.

Dist.: Malesia to Australasia and Polynesia. In New Guinea NW, E and SE.

Ecol.: Scrambling up to 10 m in forest patches. at 2100 m. At higher altitudes in moss forests, scrambling. 2100-2550 m.

Specimens: ARFAK MTS.: *Beccari s.n.* (FL); *Kostermans* 2084; *Kostermans* 2136. Hatam - *Beccari s.n.* (FL.). ANGGI LAKES: *Gibbs* 5954 (BM); *Gibbs* 5982 (BM). WITHOUT LOCALITY: *van Royen & Sleumer* 7999.

SELAGINELLACEAE

A family of some 700 species tropical, subtropical to temperate. In Papuasia some 100 species.

SELAGINELLA Pal.-Beauv

Alston, A.H.G. (1939). Notes on *Selaginella*. X. The species of the Solomon Island. Journ. Bot. London, 77: 221 - 224. f. 620.

Alston, A.H.G. (1940). The *Selaginella* of the Malay Islands. III Celebes and the Moluccas. *Bull. Jard. Bot. Btzg. III* **16**: 343-350.

Baker, J.G. (1883). A synopsis of the genus *Selaginella. Jour. Bot.* **21**: 1-5, 42-46, 80-84, 97-100, 141-145, 210-213, 240-24, 332-336.

Baker, J.G. (1887). Handbook of the Fern-Allies. A synopsis of the genera and species of the natural orders Equisetaceae, Lycopodiaceae, Selaginellaceae, Rhizocarpaceae. London.

Hieronymus, G. (1901). Selaginellaceae in Engler and Prantl, *Nat. Pfl.-fam.* **1**(4): 206, 209, 210, 621-715.

Hieronymus, G. (1913). Beitrage zur flora van Papuasien II. 10. New *Selaginella*-Arten Papuasiens nebst allgemeinen Bemerkungen uber das Vorkommen der Selaginaeelen in Papuasien. *Bot. Jahrb.* **50**: 1-45.

Jermy, A.C. (1986). Subgeneric names in *Selaginella. Fern Gaz.* **13**: 117-118.

Jermy, A.C., Jones, K. & Colden, C. (1967). Cytological variation in *Selaginella. J. Linn. Soc. Bot.* **60**: 147-158.

Spring, A.F. (1850). Monographie Famille des Lycopodiacees, Second partie. *Mem. Acad. Sci. Belg.* **24**: 1-358.

Thomas, B.A. & Quansah, N. (1991). The palaeobotanical case for dividing *Selaginella* (Selaginellaceae). *Fern Gaz.* **14**(2): 59-64.

Warburg, O. (1900). Monsunia, Beitrage zur Kenntniss der Vegetation des Sudund Ostasiatischen Monsungebletes. **1**: I-VIII, 1-208, 11 t. (1900). (*Selaginella* pp. 100-136, t. III-IV).

Selaginella angustiramea Muell. & Bak. J. Bot. 26: 26, no. 229. (1888). Gepp. in Gibbs (1917: 70); Hieron. (1910: 4); Hieron. (1913: 2); Hieron in Brause (1920: 223); Reed 55 (1965-1966); K.Sch. & Lauterbach (1905: 51); Hieron. Nat. Pflanzen.fam.682 no. 121 (19). **Type**: W. Sayer. 1887. Mt. Obree. 2100 m. (K!).

Distr.: New Guinea. Bismarck Ranges, 1800 m. 1/1902 (Schlechter 14035); Konstantinhafen. 3/1902. (Schlechter 14265). Endemic.

Ecol.: On rotting logs at 1500 m. Wet shaded rocks near waterfall at 300 m. On cliff faces near stream. Terrestrial in oak forest on mossy ground at 1400 m. Tufted pendent epiphyte on mossy tree in moist ravine at 1370 m, on mossy rocks in rainforest at 400m, on mossy rocky stream banks at 1500 - 1600 m. Epiphyte on bsae of tree trunk at 350 m in dense secondary forest. Common amongst rocks along stream on stony soil at 150m. Growing in patches around forest edges. 2100 m.

Specimens: ANGGI LAKES: *Gibbs* 5762 (BM). or 5726.

Selaginella* cf. *S. atroviridis (Wall.) Spring Flora 21: 183 (1838). Alston (1934: 53); Alston Bull. Jard. Bot. Buit. III, 14: 175 (1937); Alston Suppl. Dict. Gard. 317 (1956); Reed 62-63 (1965-1966); Ridley (1916) 265; Spring Bull. Acad. Brux. 10: 143, no. 60 (1843); Spring Monogr. Lyc. II: 124(1850); Warb. (1900) 103.

Lycopodium atro-viride Wall. (1829)

Lycopodium atro-viride Hooker & Greville (1827) [as 1831 in Reed]. **Type**: Wallich, probably Prince of Wales Isl.

S. intermedia (Blume) Spring

Dist.: India, Siam, China, Formosa, Malay Penin., Sumatra, Java, Borneo, Celebes to SW New Guinea.

Specimens: WITHOUT LOCALITY: *Croxall* 3; *Croxall* 4.

Notes: Reed as `*atrovirens*'.

Selaginella canaliculata (L.) Bak. var. ***d'urvillei*** (Bory) Warb. *Monsunia* **1**: 107, 122 (1900).

Lycopodium durvillaei Bory (1828). **Type**: ?Warburg 12556. Middle Luzon etc. Reed 77 (1965-1966); K.Sch. & Lauterbach (1905) 152.

Dist.: E. Malesia, Papuasia to Melanesia. In Papuasia: Aru Is., Irian Jaya, Bism. Arch.

Specimens: ARFAK MT.: Putat - *Beccari s.n.* (FL).

Notes: Warburg & Lauterbach (1905) as *S. durvillaei* [see entry].

Selaginella caulescens (Wall.) Spring *Bull. Acad. Brux.* **10**: 137, no. 12 (1843). Alston Bull. fam. Mem. Inst. 5: 285 (1934); Alston Bull. Jard. Bot. Buit. III, 13: 441 (1935); Alston Bull. Fan. Mem. Inst. III, 14: 175, 185 (1937); Baker 24 (1885); Baker Fern Allies 94, no. 232 (1887); Beccari 55 (1886); Beccari in d'Albertis 400 (1880); F. Muell. 75 (1876); F. Muell. 20 (1885); Reed 79 (1965-1966); Spring Monogr. Lyc. II: 158 (1850); Warb. 107, 121 (1900).

Lycopodium caulescens Wall. (1829) [nomen], ex Hook. & Grev. (1831). **Type**:

Selaginella involvens (Swartz) Spring [Alston 347 (1940)].

Dist.: S and E Asia, Sumatra, Java, Philippines to New Guinea. In Papuasia: W, SE New Guinea.

Specimens: RAMOI: *Beccari s.n.* (FL). PUTAT: *Beccari s.n.* (FL).

Selaginella* c.f. *S. muelleri Baker *J. of Bot.* **23**: 122 (1885). Baker, Fern Allies 94 no. 231 (1887); Hieron. 2 (1913); Hieron. Nat. Pflanzen.fam. 682 no. 120; F. Muell. 20 (1885); Reed 162 (1965-1966); K.Sch. & Lauterbach 50 (1905); Wagn. & Greth. 64 (1948); Warb. 104 (1900). **Type**: Edelfeldt 138, 159. 1884. Near Port Moresby; Armit s.n. 1884. From Islands near the South East Coast of New Guinea. (K! Syntypes) [see Reed 162].

Dist.: NE SE New Guinea. Bismarck Range, 1800m; Simbang, 100m (Nymann 792. 8.1899); Admiralty Islands.

Ecol.: On shady creek bank at 50 m, gregarious in rocky ground at 500 m, along river banks and in forest at 15 m to 450 m.

Specimens: WITHOUT LOCALITY: *Croxall* 2.

Selaginella* cf. *S. velutina Cesati *Rendic. Accad. Sc. Fis. Mat. Napoli* **16**: 28, 31 (1877). **Type**: Beccari s.n. ex Ramoi maggio 1872. (FL). Alston (1940: 350); Beccari (1886); Brass (1941: 291); Hieron. (1913: 2); Hieron in Brause (1920: 239); Reed (1965-1966: 232).
Selaginella wallichii (Beccari,1886)
Selaginella motiensis Hieron 1901. (Ternate & Moti Isl.)

Dist.: Moluccas to NW New Guinea.

Specimens: WITHOUT LOCALITY: *Croxall* 0.

Selaginella vonroemerii Alderw. *Bull. Jard. Bot. Buit.* II **24**: 7 (1917). Reed 235 (1965-1966). **Syntypes**: von Roemer 779, Pulle 531 (L).

Dist.: Irian Jaya -

Specimens: WITHOUT LOCALITY: *van Royen & Sleumer* 5066.

Selaginella wallichii (Hook. & Grev.) Spring Baker (1876: 112); Baker (1885: 19); Muell. (1885a: 20); Muell. (1885b: 20); Reed (1965-1966: 236).
Lycopodium wallichii Hook. & Grev. ex Grev. & Hook. (1831). **Type**:

Dist.: E Himalayas, Malay Peninsula, Philippines to New Guinea. In Papuasia: NW SE New Guinea.

Specimens: RAMOI: *Beccari s.n.* (FL).

Notes: Warburg (1900) states that the record from NG is doubtful.

***Selaginella* spp.**

Specimens: ARFAK PLAINS: *Johns* 8151. S. Wariori Crossing - *Sands* 6282. ANDAI: *Johns* 8109. ARFAK MTS.: Mokwam Trail - *Johns* 7690; *Johns* 7746; *Johns* 7750. Mupi Dessa - Sands 6829; Sands 6883. ANGGI LAKES: *Johns* 7966. KEBAR VALLEY: Andjai 2 - *Johns* 8250; *Johns* 8296; *Johns* 8318; *Johns* 8340; *Johns* 8354. Apni Valley - *Johns* 8252. K.Anjai 1 to K.Akmuri - *Johns* 8335. IRAY: *Johns* 7977; *Johns* 7987. G. MEJA: *Johns* 7650; *Johns* 7652. RANSIKI: *Johns* 7853; *Johns* 7868. AREA 12: *Johns* 8383. AREA 18: *Johns* 8384. WITHOUT LOCALITY: *Croxall* 6; *van Royen & Sleumer* 6865; *van Royen & Sleumer* 7721.

FERNS

R. J. JOHNS, P. J. EDWARDS and D. KOWALSKA

ADIANTACEAE

CHEILANTHES L.

Cheilanthes tenuifolia (Burm. f.) Sw. *Syn. Fil.* 129: 332 (1806). Copeland (1949: 36); F.Muell. (1875: 48).
Trichomanes tenuifolium Burm. f. *Fl. Ind.* 237 (1768).
Dist.: S.E. Asian Tropics. Throughout New Guinea to Solomons, into the Pacific.
Ecol.: Herbaceous plant in dryland areas such as grassland at 520 m altitude.
Specimens: KEBAR VALLEY: SW Andjai - *van Royen & Sleumer* 6725 (K, L).
Notes: Leaves brownish-green.

***Cheilanthes* sp.**
1. **Ecol.**: Terrestrial in open forest near hot springs.
Specimens: KEBAR VALLEY: Andjai 2 - *Johns* 8317.
Notes: Fronds dimorphic. Sterile fronds: stipe to 4-5 cm, rachis to 8-10 cm. Fertile fronds: stipe to 25 cm, rachis to 18 cm. Pinnae margins recurved.

ONYCHIUM Kaulf.

Onychium siliculosum (Desv.) C.Chr. *Ind. Fil.* 20 (1905). Alderw. (1908: 330-331); C.Chr. (1906: 469); Copeland (1949: 39).
Pteris siliculosa Desv. *Berlin Mag.* **5**: 324 (1811).
Onychium auratum Kaulf. *Enum.* 144 (1824).
Onychium tenue Christ *Bull. Herb. Bois.* II **1**: 451 (1901).
Dist.: Tropical Asia to New Guinea. Widespread in New Guinea up to an altitude of 1000 m.
Ecol.: Terrestrial, growing singly or in small clumps usually in gullies. In PNG common on landslides. In Vogelkop on conglomerate bank in open.
Specimens: ARFAK MTS.: *O. Beccari* s.n. (FI, K). RANSIKI RD.: Near Kali Urkwan between Wanmarwai and Warkapi - *Sands* 6693.
Notes: Plants to 50 cm. Reverse side of fronds yellowish-green.
Clemens 41935 was collected by 'native collectors' at 1500-1800 m. The altitude is presumably not correct.

PITYROGRAMMA Link.

Pityrogramma calomelanos (L.) Link *Handb. de Gew.* **3**: 20 (1883).
Acrostichum calomelanos L. *Sp. Pl.* **2**: 1072 (1753).
Dist.: Introduced.
Ecol.: Common in disturbed areas, along road sides through disturbed lowland rain forest. In open gravel of flood plain. Altitudinal range 50-650 (-1600) m.
Specimens: ARFAK PLAINS: Wariori River - *Johns* 8147. RANSIKI RD.: *Johns* 7903; *Sands* 6214. ARFAK MTS.: Mupi Dessa Trail - *Sands* 6739.

Notes: Reverse of fronds silvery.
Widely used for facial decoration in New Guinea.

SYNGRAMMA J.Smith

Syngramma schlechteri Brause 52 (1912). **Type**: R. Schlechter 18164. Finisterres Ranges, Papua New Guinea. 1000 m. (B).

Craspedodictyum schlechteri (Brause) Copel. 38 (1929).

Syngramma grande (Copel.) C.Chr. *Ind. Fil. Suppl.* 17 (1913). Copeland (1936: 115); Copeland (1949: 26).

Craspedodictyum grande Copel. 84 (1911). **Type**: L.J. Brass 2865. Hinuahaoro, San Cristoval, Solomon Is. 900 m. (BO: K).

Dist.: New Guinea to Solomon Islands.

Ecol.: Terrestrial on very steep slopes in hill forest dominated by *Lithocarpus* and *Hopea*. Particularly common in montane forest and *Nothofagus* forest. 550 - 1000 m altitude.

Specimens: ARFAK PLAINS: On steep ridges to G. Itsiwei - Sands 6430.

Notes: Sterile fronds to 30 cm. Fertile fronds to 55 cm. Stipe dull green, maroon at base.

***Syngramma* spp.**

1. **Ecol.**: Terrestrial in disturbed lowland rain forest.
Specimens: ARFAK PLAINS: Masni - *Johns* 8189;
Notes: Bulbills at end of pinnae.

2. **Ecol.**: Terrestrial in shade on forest floor amongst leaf litter in partially disturbed lowland rain forest at 0 - 20 m altitude.
Specimens: ARFAK PLAINS: Sungai Wariori -*Sands* 6265.
Notes: Fronds to 50 cm long. Stipes dull green. Margins of some fronds crenate.

TAENITIS Willdenow ex Spregel

Taenitis blechnoides (Willd.) Swartz *Syn. Fil.* **24**: 220 (1906).

Pteris blechnoides Willd. *Phytogr.* **12**: 9. t.3 (1794). **Type**: Klein. S India. (Holotype: Herb. Willd. B).

Dist.: From Sri Lanka and S India, throughout SE Asia and Malesia to Australia and the Caroline Islands and Fiji. Widespread in New Guinea. Common in Irian Jaya with many collections from 20-1450 m.

Ecol.: Very common terrestrial of mixed lower montane forest dominated by *Castanopsis* and *Araucaria* mixed with *Lithocarpus* and *Litsea*. Also common in lowland rain forest. Sea-level to 1500 m altitude.

Specimens: ARFAK MTS. : Mokwam Trail - *Johns* 7809. GEELVINK BAY: *Schonian* 83 (B).

Taenitis hookeri (C.Chr.) Holttum *Blumea* 91 (1968). Copeland (1949: 25).

Syngramma hookeri C.Chr. Index (1905: 346). C.Chr. (1906: 629).

Taenitis lanceolata (Diels) Holttum *Blumea* 91 (1968) not ***T. lanceolata*** (L.) Kaulf.

Hemionitis lanceolata Hooker. **Type**: Milne. H.M.S. Herald. Fiji. 180 m. (K).

Dist.: Borneo to New Guinea, the Solomon Islands and Fiji.

Ecol.: Terrestrial in montane forest. Common along ridges in forest dominated by *Castanopsis* and *Lithocarpus*, alos along creeks and ravines. Abundant on lateritic soils near Sorong. 600-1750 m altitude. Also in disturbed forest in shade near hot springs at 625 m alt.

Specimens: ARFAK PLAINS: On steep ridge to S along track to G. Itsiwei - Sands 6401. RAMOI: Beccari s.n. 1872 (FI). KEBAR VALLEY: Bukit Moropim - *Sands* 6484; Hot Springs - *Sands* 6628.
Notes: Fronds to 30 cm long, dimorphic.

ASPLENIACEAE

Copeland, E.B. (1949, 1950). Aspleniaceae and Blechnaceae of New Guinea. *Philip. J. Sci.* **78**: 207-229.

Kato, M. (1992). Taxonomic studies of pteridophytes of Ambon and Seram (Moluccas) Collected by Indonesian-Japanese Botanical Expeditions VIII. Aspleniaceae, Blechnaceae. *J. Fac. Sci. Univ. Tokyo, Sect.* III **15**: 111-133.

ASPLENIUM L.

Asplenium acrobryum Christ *Nova Guinea* **8**: 150 (1909).
Notes: Study of the spores of this species indicates that the Type specimen is of hybrid origin. The following listed specimens will need detailed study before the new species are described.

Asplenium acrobryum **complex sp. 1**
Ecol.: Climbing on tree trunk with short-creeping rhizomes in disturbed lowland rain forest. 0-50 m.
Specimens: ARFAK PLAINS: Masni - *Johns* 8168;
Notes: Fronds nonsoriferous in either basal or apical sections.

Asplenium acrobryum **complex sp. 2**
Ecol.: Epiphyte in shade in disturbed lowland rain forest at 0-20 m altitude.
Specimens: ARFAK PLAINS: Near road to S. Wariori - *Sands* 6250.
Notes: Fronds to 5 cm (?), rachis dark brown.

Asplenium acrobryum **complex sp. 3**
Ecol.: In disturbed forest near Hot Springs.
Specimens: KEBAR VALLEY: Hot Springs - *Johns* 8377.
Notes: Rhizome short-creeping with scattered long brown scales.

Asplenium acrobryum **complex sp. 4**
Ecol.: In disturbed lowland rain forest on Wariori flood plain at c. 50 m altitude.
Specimens: ARFAK PLAINS: Masni - *Johns* 8176.
Notes: Fronds nonsoriferous in either basal or apical sections.

Asplenium acrobryum **complex sp. 5**
Ecol.: In disturbed lowland rain forest on Wariori flood plain at c. 50 m altitude.
Specimens: ARFAK PLAINS: Masni - *Johns* 8171.
Notes: Stipe black to 2 cm long with numerous brown scales. Fronds to 82 x 8.6 cm, densely soriferous.

Asplenium acrobryum **complex sp. 6**
Ecol.: Epiphyte with creeping rhizome in lower montane forest dominated by *Araucaria*, *Lithocarpus* and *Litsea*. c. 650 m altitude.
Specimens: ARFAK MTS.: Mockwam Track - *Johns* 7793.

Notes: Stipe black to 2.5 cm long with large scales. Fronds to 50 x 5.5 cm. Soriferous to apex of frond..

Asplenium acrobryum complex **sp. 7**
Ecol.: Epiphyte with creeping rhizome in lower montane forest dominated by *Araucaria*, *Lithocarpus* and *Litsea*. c. 650 m altitude.
Specimens: ARFAK MTS.: Mockwam Track - *Johns* 7807.
Notes: Stipe to 1 cm long, with long brown scales. Fronds to 50 x 4.5 cm.

Asplenium acutiusculum Bl. *Enum*. 178 (1828). Gibbs (1917: 73).
Dist.: Throughout Malesia - Java, Celebes, Philippines, New Guinea to Samoa.
Ecol.: Epiphytic in forest patches. 2100 m.
Specimens: ANGGI LAKES: *Gibbs* 5730 (BM).
Notes: Copeland (1950) states that this species is not recorded from New Guinea.

Asplenium aff. ***A. affine*** Sw. *Schrad. Journ.* **1800**(2): 56 (1801). Beccari (1886: 40); Copeland (1911: 79); Copeland (1950: 220); Copeland (1961: 445); Hieron. in Schumann (1893: 84); Hieron. in Brause (1920: 149); Kato (1992: 143); Ridley (1916: 256); Schumann & Lauterbach (1900: 129); Sledge (1962: 401); Sledge (1965: 269); Tagawa & Iwatsuki (1985: 288). **Type**:
Dist.: Tropical Africa (Madagascar) to Sri Lanka, S India, Indo China, throughout Malesia to Polynesia (New Hebrides & Fiji).
Ecol.: In secondary forest on uplifted coral along road side at 110 m altitude.
Specimens: GUNUNG MEJA: *Johns* 7658.
Notes: Rhizome short-creeping. Vascular tissue arranged in a circle in 4 segments. Bulbills rare near apex of some fronds.
Sledge (1962) includes extensive notes on this species.

Asplenium amboinense Willd. *Sp. Pl.* **5**: 303 (1810). Mettenius n. 15. Alderw. (1908: 441); Alderw. (1916: 109); Beccari (1886: 40); Brause (1920: 145); Cesati (1877: 26); Copeland (1950: 78); Lauterbach (1929: 568); Schumann & Lauterbach (1905: 43); Wagner & Grether (1948: 56). **Type**:
Dist.: Malesia (Java, Borneo, Sulawesi, Ambon and Ceram and New Guinea) to Polynesia.
Specimens: It is possible that some of the collections listed under *A. acrobryum* belong here.
Notes: A difficult complex in much need of detailed study. Specimens are often confused with *A. acrobryum* and several other undescribed species.

Asplenium beccarianum Cesati
Asplenium bipinnatifidum Baker var. ***beccarianum*** (Cesati) C.Chr.

Asplenium belangeri Kunze *Bot. Zeit.* **1848**: 176 (1848).
Dist.: India to New Guinea.
Ecol.: Epiphyte in hill forest. 600 - 1000 m altitude.
Specimens: KEBAR VALLEY: *van Royen & Sleumer* 6921 (L).

Asplenium bipinnatifidum Baker *Syn.* **1**: 221 (1867).
Asplenium cesatianum Baker *Malesia* **3**: 39 (1886). Brause (1920: 145). **Type**: Beccari s.n. Putat. Arfak Mts. (FI).
Dist.: New Guinea to Pacific.
Ecol.: Epiphyte in lower montane forest.
Specimens: WITHOUT LOCALITY: *Beccari s.n.* (FI); *Beccari s.n.* (FI).

Asplenium bipinnatifidum Baker var. ***beccarianum*** (Cesati) C.Chr. *Brittonia* **2**: 289 (1937). Beccari (1886: 41); Hieron. in Brause (1920: 146); Kato (1992: 143). **Type**: Beccari s.n. Hatam, Mt Arfak. (FI).

Asplenium beccarianum Cesati *Rend. Acad. Napoli* **16**: 26, 29 (1877).

Dist.: New Guinea to Ceram.

Ecol.: Very common trailing epiphyte in lower montane forest dominated by *Araucaria*, *Lithocarpus* and *Litsea* at 650 m altitude.

Specimens: ARFAK MTS.: Hatam - Beccari s.n. (FI); Mokwam Trail - *Johns* 7786.

Notes: The collections from New Guinea have distinctly larger leaves with broader segments and less finely dissected, smaller fronds. Rachis usually short to 2-3 cm. Pinnae short to 2 cm long. 6-8 pinnae per frond.

Asplenium caudatum Forst. *Prod.* 80 (1786). Mettenius no 139. Copeland (1961: 435); Hieron. in Brause (1920: 149); Holttum (1955: 247); Lauterbach (1929: 6); Ridley (1916: 256); Schumann & Lauterbach (1905: 128).

Dist.: SE Asian to Polynesia.

Ecol.: Epiphyte in montane forest from 1000 - 2000 m altitude.

Asplenium cuneatum Lam. *Enc.* **2**: 309 (1786). Mettenius no 72. Baker (1876: 107); Carr. in Seem. (1873: 355); Copeland (1911: 79); Copeland (1950: 220); Hieron. in K.Schumann & Lauterbach (1898: 84); Kuhn (1889: 7); Lam (1828: 281); Lam (1945: 32); Lauterbach (1929: 566); Lauterbach (1929: 23); Lauterbach (1930: 431); Wagner & Grether (1943: 56).

Dist.: SE Asian.

Asplenium cyatheifolium Bory. Beccari (1886: 41); Cesati (1877: 26); Miquel (1869: 169); Mueller (1876: 52). **Type**: d'Urville. New Guinea. (P).

See *Cheilanthes tenuifolia*

Dist.: Endemic. W New Guinea.

Specimens: ANDAI: *Beccari s.n.* (FI). ARFAK MTS.: Putat - *Beccari s.n.* (FI).

Notes: This is probably a species of *Diplazium*.

Asplenium laserpitiifolium Lam. *Ency.* **2**: 310 (1786). Mettenius no 163. Baker (1890: 106); Beccari in d'Albertis (1880: 399); Beccari (1886: 40); Carr. in Seem. (1873: 355); Cesati (1877: 26); Copeland (1929: 271); Gepp in Gibbs (1917: 195); Gepp in Rendle (1923: 60); Hieron. in Brause (1929: 150); Kuhn (1889: 7); Lauterbach (1929: 8); K.Schumann & Lauterbach (1900: 129).

Dist.: Moluccas, New Guinea to Polynesia and tropical Australia.

Ecol.: Common near sea level.

Specimens: RAMOI: *Beccari s.n.* (FI). WITHOUT LOCALITY: *van Royen & Sleumer* 7031 (L).

Notes: Presl (1849) treated this species in the genus *Tarachea*.

Asplenium lobulatum Mett. ex Kuhn *Linn.* **36**: 100 (1869). Hooker and Baker 486.

Dist.: New Guinea to Hawaii.

Specimens: ARFAK MTS.: *Croxall* 8.

Asplenium macrophyllum Swartz *Schrad. Journ.* **1800**(2): 52 (1801). Holttum (1955: 249 as *A. falcatum*); Copeland (1961: 437); Kato (1992: 141). **Type**:

Dist.: India and Ceylon, throughout Malesia to Polynesia.

Ecol.: Epiphyte on tree trunk of *Ficus*, 3-4 m above ground.

Specimens: RANSIKI RD: *Johns* 7812.
Notes: Rhizome short creeping, scales dense, golden brown. Stipe black to 25 cm long. Fronds pinnate with 6-9 pinna, apical pinna tri-lobed.

Asplenium polypodioides Swartz *Schrad. Journ.* **1800**(2): 53 (1801).
= ***A. platyneuron***
Specimens: ARFAK MTS. : *Beccari s.n.* (FI).

Asplenium robustum Blume *Enum.* 189 (1828). Copeland (1961: 446); Holttum (1955: 442 as *laserpitifolium*); Kato (1992: 144). **Type**:
Dist.: W Malesia. Borneo, Philippines, Sulawesi, Moluccas ? to New Guinea.
Ecol.: Epiphyte, rarely epipetric or terrestrial in lowland or montane forest.
Specimens: *van Royen & Sleumer* 4991.
Notes: **? = A. laserpitiifolium**

Asplenium scandens J.Sm. ex Mett. *Fl. Lechl.* **2**: 108 (1857). J. Sm. *nomen. J. of Bot.* **3**: 408 (1841). Copeland (1961: 443); Gepp in Gibbs (1917: 73); Mettenius Aspl. 108 n. 57 (1859).
Darea scandens (J.Sm.)
Dist.: Borneo, Philippines, Sulawesi, Moluccas, New Guinea and Solomon Islands to Fiji.
Ecol.: Epiphyte, epipetric or terrestrial in lowland or montane forest to 2100 m altitude.
Specimens: ANDAI: *Beccari s.n.* (FI). ANGGI LAKES: *Gibbs* 5642 (BM). RAMOI: *Beccari s.n.* (FI).

Asplenium spinulosum Mettenius *Aspl.* 172, no. 189 (1859).
= ? ***Diplazium proliferum***
Specimens: ANDAI: *Beccari s.n.* (FI).

Asplenium sylvaticum (*vide* ***silvaticum***) Presl *Rel. Haen.* **1**: 42 (1825).
= ? ***Diplazium***
Specimens: RAMOI: *Beccari s.n.* (FI).

Asplenium tenerum Forst. *Prod.* 80 (1786). Mettenius n. 64. Copeland (1961: 431); Holttum (1955: 432); Kato (1992: 142); Tagawa & Iwatsuki (1985: 273).
Asplenium tenerum Forst. var. ***acuminatum*** (Hook.) Alderw. Handb. Mal. Ferns 458 (1908).
Dist.: India and Ceylon, throughout Malesia to Polynesia.
Ecol.: Usually epiphytic rarely terrestrial in lowland and lower montane forest.
Specimens: ANDAI: *Beccari s.n.* (FI). ARFAK MT.: Putat - *Beccari s.n.* (FI).
Notes: ? = ***A. caudatum*** Cav. (1802) *non* Forst. (1786).

Asplenium thunbergi Kunze *Linn.* **10**: 517 (1836). Holttum (1968: 634); Kato (1992: 143); Tagawa & Iwatsuki (1985: 274).
Asplenium doreyi Kunze. Beccari (1886: 40); Cesati (1877: 26). **Type**: Dorei. New Guinea.
Asplenium decorum Kunze. Copeland (1950: 214); Sledge (1956: 276).
Dist.: Sri Lanka, Indochina, throughout Malesia.
Ecol.: In distured hill forest near stream in partial shade. Epiphyte on trunk of fallen tree. 395 m alt.
Specimens: ARFAK PLAINS: *Sands* 6264; Beccari s.n. (FI). ARFAK MTS.: Mupi Dessa trail (Kali Meniyepto Valley) - *Sands* 6876.
Notes Stipe light green, fronds to 30 cm.

Kunze *Anal.* 23 (1837) = *A. tenerum*

***Asplenium* spp.**

1. **Specimens**: *Mayar* 398.

2. **Specimens**: *Sands* 6129.

3. **Specimens**: *Sands* 6126.

4. **Ecol.**: In riverine forest in deep shade as epiphyte on old tree trunk in lower montane zone. 450 m alttiude.
Specimens: ARFAK MTS.: Mupi Dessa Trail to G. Humibou - *Sands* 6725.
Notes: Erect epiphytic fern to 60 cm tall. Frond with 20-30 pinnae. Stipe green.

5. **Ecol.**: In riverine forest in deep shade in lower montane zone. 450 m alttiude.
Specimens: ARFAK MTS.: Mupi Dessa Trail to G. Humibou - *Sands* 6726.
Notes: Plants with bulbills near apex of frond (1 m long). Stipe dark maroon.

6. **Ecol.**: Climbing on tree trunk over river in shade. 30-50 m alt.
Specimens: ARFAK PLAINS: In river valley on southern margin - *Sands* 6339.
Notes: Rhizome stout, climbing. Fronds to 80-90 cm long with stipes to 15-24 cm long.

7. **Ecol.**: Climbing on fallen tree trunk in shade in lowland rain forest. 30-50 m altitude.
Specimens: ARFAK PLAINS: Foothills to south - *Sands* 6338.

8. **Ecol.**: Climbing on fallen trunk in lowland rain forest in shade.
Specimens: ARFAK PLAINS: Foothill forests along southern margins - *Sands* 6338.
Notes: Rhizome flattened, fronds to 45 cm long. Pinnae 7-8 pairs and terminal one.

9. **Ecol.**: Epiphyte in montane forest in deep shade at 630 m.
Specimens: ARFAK MTS.: Mupi Dessa Trail - *Sands* 6858.
Notes: Fronds to 70 cm long.

10. **Ecol.**: Epiphyte on trunk in shade in montane forest at 850 m altitude.
Specimens: ARFAK MTS.: Mupi Dessa Trail - *Sands* 6818
Notes: Epiphyte with erect fronds to 75 cm long. Tufted from short rhizome.

11. **Ecol.**: Epiphyte in shade in lower montane forest at 875 m altitude.
Specimens: ARFAK PLAINS: On ridge to G. Itsiwei - *Sands* 6396.
Notes: Epiphyte with spreading fronds variable in length to 40-150 m.

12. **Ecol.**: Epiphyte in ridge forest in lower montane forest from 600-650 m altitude.
Specimens: ARFAK MTS.: Mokwam Trail - *Johns* 7729.

13. **Ecol.**: In forest at 70 m altitude.
Specimens: AREA 18: *Johns* 8382.
Notes: Stipes to 28 cm long. Fronds pinnate, rachis to 32-50 cm long.

14. **Ecol.**: Terrestrial in ridge top forest along limestone escarpment at 900 m altitude.
Specimens: KEBAR VALLEY: E of Andjai 2 - *Johns* 8268.

Notes: Rhizome clumped. Stipes black, to 40-44 cm, with dense brown scales at base. Rachis to 30 cm. Sori scattered.

15. **Ecol.**: Epiphyte on *Pandanus* trunk in ridge top forest along limestone escarpment at 900 m altitude.
Specimens: KEBAR VALLEY: E of Andjai 2 - *Johns* 8269.
Notes: Stipes blackish, to 25-27 cm long. Fronds pinnate, rachis to 35 cm.

16. **Ecol.**: Epiphyte in ridge top forest along limestone escarpment at 900 m altitude.
Specimens: KEBAR VALLEY: E of Andjai 2 - *Johns* 8271.
Notes: Stipes green, to 13-18 cm.

17. **Ecol.**: Epiphyte in ridge top forest along limestone escarpment at 900 m altitude.
Specimens: KEBAR VALLEY: E of Andjai 2 - *Johns* 8273.
Notes: Rhizome forming a small caudex, stipes black, to 16 cm, rachis to 35 cm.

18. **Ecol.**: In partially disturbed lowland rain forest at 0-20 m altitude.
Specimens: ARFAK PLAINS: Wariori crossing - *Sands* 6249.
Notes: Fronds to 1 m long and 15 cm wide.

19. **Ecol.**: Epiphyte in ridge forest in lower montane forest from 600-650 m altitude.
Specimens: ARFAK MTS.: Mokwam Trail - *Johns* 7730.
Notes: Contrasts with *Johns* 7729 in the presence of copious black clathrate scales at the base of long stipes (to 16 cm).

20. **(Loxoscaphe)**
Ecol.: Climbing rhizome with fronds to 2.3 cm apart. Valley floor forest in lower montane forest from 600-650 m altitude.
Specimens: ARFAK MTS.: Mokwam Trail - *Johns* 7752.

21. **Ecol.**: Epiphyte on tree trunks in disturbed lowland rain forest at c. 50 m altitude.
Specimens: ARFAK PLAINS: Wariori crossing - *Johns* 8161.
Notes: Fronds bipinnate. Stipe black.

22. **Ecol.**: Low epiphyte on tree trunks in ridge top forest in lower montane forest dominated by *Araucaria*, *Lithocarpus* and *Litsea* at 650 m altitude.
Specimens: ARFAK MTS.: Mokwam Trail - *Johns* 7780.
Notes: Climbing rhizome with long pointed, greyish-brown scales. Stipes to 1 cm long. Fertile fronds to 48 cm long, densely soriferous.

23. **Ecol.**: Climbing on stumps in shade on very steep ridge in montane forest at 1050 m altitude.
Specimens: ARFAK MTS.: Mupi Dessa Trail to G. Humibou - *Sands* 6757.
Notes: Climber to 1.5 m.

ASPLENIUM Sect. THAMNOPTERIS

Johns, R.J. (1996). A field study of *Asplenium* sect. *Thamnopteris* in the NE Kepala Burung, Irian Jaya, Indonesia. In: Camus, J.M., Gibby, M. and Johns, R.J. (eds.). *Pteridology in Perspective*. pp. 337-340. R.B.G. Kew.

Asplenium nidus L. *Sp. Pl.* 1079 (1753). Baker (1876: 106); Beccari (1886: 39); Brass (1941: 295); Bonap. (1921: 231); Burk. (1896: 92); Carr. in Seem. (1873: 353); Cesati (1877: 25); Copeland in C.T. White (1929: 271); Copeland (1950: 209); Dakkus (1930: 33); Gaud. (1828: 313); Hieron. in K.Schumann (1898: 84); Hieron. in Brause (1920: 144); Lam (1924: 5); Lam (1945: 45); Lauterbach (1929: 10); Lauterbach (1930: 445, 452-457); Mettenius 91866: 233); mIQ. (1869: 163); Schlechter in Brause (1912: 2,4,6); K. Schumann & Lauterbach (1900: 126); Sledge (1965: 242); Wagner & Grether (1948: 55). **Type**: Osbeck s.n. Java. (Herb LINN).

Dist.: Old World Tropics.

Ecol.: Epiphyte from coastal to ridge forest, 0 - 1000 m.

Specimens: ARFAK MTS.: *Johns* 8102; *Johns* 8133.

Notes: Specimens provisionally identified as *A. nidus* L. (not confirmed) include: *Beccari s.n.* (FI). ARFAK MTS.: Muppi Dessa - *Sands* 6730.

A revision of this section is urgently required as several undescribed species occur in this group throughout its range.

***Asplenium* sp. nov. A**

Ecol.: Epiphyte in rain forest.

Specimens: ARFAK PLAINS: Wariori Flood Plain - *Johns* 8229; *Johns* 8230; *Johns* 8246; *Johns* 8247.

Notes: Rhizome to 10-25 cm long. Scales dense, black. Stipe short to 2.5-3 cm. Fronds to 78-98 cm long. See tables with collections - *Johns* 8229, *Johns* 8230, *Johns* 8246, *Johns* 8247.

***Asplenium* sp. nov. B**

Ecol.: Epiphyte in primary rain forest at 60-70 m altitude.

Specimens: ARFAK PLAINS: Wariori Flood Plain - *Johns* 8231.

Notes: Distinct, very fine scales on stipe. Sori very short. See tables with collection - *Johns* 8231.

***Asplenium* sp. nov. C**

Ecol.: Terrestrial on limestone and near hot springs in Kebar Valley from 600 - 650 m altitude.

Specimens: KEBAR VALLEY: On limestone N of Andjai - *Johns* 8360. Near Hot Springs - *Johns* 8367.

***Asplenium* sp. nov. D**

Ecol.: Epiphyte in primary rain forest at 60-70 m altitude.

Specimens: ARFAK PLAINS: Wariori Flood Plain - *Johns* 8234; *Johns* 8235; *Johns* 8236, *Johns* 8248.

Notes: Rhizomes to 25-33 cm long. See tables with collections - *Johns* 8234, *Johns* 8235, *Johns* 8236, *Johns* 8248.

***Asplenium* section *Thamnopteris* spp.**

1. **Ecol.**: Epiphyte on fallen trunk on lower steep slopes. In lower montane forest dominated by *Hopea* and *Lithocarpus*.
   Specimens: ARFAK PLAINS: Along steep ridges to G. Itsiwei - *Sands* 6435.
   Notes: Erect ferns with lanceolate fronds t 30-40 cm long. Stipe short, maroon. Main veins maroon.

2. **Ecol.**: On rocks and fallen tree trunks close to stream in shade. In montane forest at 920 m.
   Specimens: ARFAK MTS.: Muppi Dessa Trail - *Sands* 6854.

Notes: Data on frond in table. Lacking stipe.

3. **Ecol.**: Epiphyte in rainforest on very dry limestone area at 100-200 m altitude.
Specimens: RANSIKI: Along road c. 5km NW of Ransiki - *Johns* 7877.
Notes: Rhizome creeping on trunk. Fertile fronds to 120x32 cm. Material in blackened bases is almost completely composed of roots - virtually no humus. See frond measurements in table on sheets.

4. **Ecol.**: Epiphyte in forest at 1700-1940 m altitude.
Specimens: IRAY: *Johns* 7996.
Notes: Rhizome to 12+ cm long x 7 cms broad. Apex covered with dense black clathrate scales.

5. **Ecol.**: Epiphyte on subcanopy trees in disturbed lowland rain forest at 50 m altitude.
Specimens: ARFAK PLAINS: Along steep ridges to G. Itsiwei - *Sands* 6435.
Notes: Nest fern with climbing rhizomes to 20 cm tall. Fronds lacking stipes. Table of data with collection.

6. **Ecol.**: Low epiphyte on tree trunks in ridge forest in lower montane forest dominated by, *Araucaria*, *Lithocarpus* and *Litsea* at 650 m alttiude.
Specimens: ARFAK MTS.: Mokwam Trail - *Johns* 7790.
Notes: Fronds many, arising almost in a rosette. Fronds to 130 cm long and 16 cm broad. Apical 10 cm of frond infertile.

7. **Ecol.**: Low epiphyte on tree trunks in ridge forest in lower montane forest dominated by, *Araucaria*, *Lithocarpus* and *Litsea* at 650 m alttiude.
Specimens: ARFAK MTS.: Mokwam Trail - *Johns* 7775.
Notes: Stipe black to 10 cm long. Fringed to 3-4 mm wide. Fronds simple to 90 x 8.3 cm. Intramarginal veins indistinct.

8. **Ecol.**: Low epiphyte on tree trunks in ridge forest in lower montane forest dominated by, *Araucaria*, *Lithocarpus* and *Litsea* at 650 m alttiude.
Specimens: ARFAK MTS.: Mokwam Trail - *Johns* 7800.
Notes: Climbing rhizome with fronds very close. Stipes to 3-4 cm long. Fronds to 92 cm long. Sori to 5-6 cm long. Spaced on every 2nd or 3rd vein.

9. **Ecol.**: Low epiphyte on tree trunks in ridge forest in lower montane forest dominated by, *Araucaria*, *Lithocarpus* and *Litsea* at 650 m alttiude.
Specimens: ARFAK MTS.: Mokwam Trail - *Johns* 7808.
Notes: Climbing rhizome with fronds spaced at 15-20 cm. Stipes black, to 2 cm long. Fronds to 180 x 11-12 cms. Sori to 1-1.5 cm long.

10. **Ecol.**: Low epiphyte on tree trunks in ridge forest in lower montane forest dominated by, *Araucaria*, *Lithocarpus* and *Litsea* at 650 m alttiude.
Specimens: KEBAR VALLEY: Kampng Akmuri - *Johns* 8332.
Notes: Stipes to 5 cm long, densely scaly at base. Fronds to 150 cm x 19-24 cm. The base forming a small cup full of roots.

***Asplenium* spp. Specimens**: ARFAK PLAINS: Wariori Flood Plain - *Johns* 8161. KEBAR VALLEY: Kampong Akmuri - *Johns* 8338. Anjai 2 - Johns 8360. RANSIKI: *Sands* 6249.

DIPLORA Baker

Diplora d'urvillaei Kunze, *Rev. Gen. Pl.* **2**: 818 (1891). Brause (1920: 144); Copeland (1913: 153).
Dist.: New Guines.
Specimens: *Beccari* 12525 (FI); *Beccari* 12526 (FI).

Diplora longifolia (Presl.) Copeland
Micropodium longifolium (Presl) Mett. *Ann. Mus. Bot. Lugd.-Bat.* **2**: 233 (18).
Scolopendrium longifolium Presl *Rel. Haenk.* **1**: 48. pl. 9. fig. 1 (18).
Dist.: Philippines and New Guinea to Carolines.
Specimens: *van Royen & Sleumer* 7309.
Notes: Copeland (1913) treated this as a synonym of *Phyllitis d'urbillei.*

***Diplora* sp. nov.**
Ecol.: Terrestrial at base of trees adjacent to hot springs at 650 m altitude.
Specimens: KEBAR VALLEY: Hot Springs - *Johns* 8314.
Notes: A new species of *Diplora.*

Stipe black to 50-60 cms long. Fronds pinnate to 40 cm long with 13 pinnae and a terminal pinna. Sori sometimes long with a short paired region.

***Diplora* spp.**

1. **Ecol.**: Climber on subcanopy trees and shrubs at 70 m..
Specimens: ARFAK PLAINS: Valley W of Wariori - *Johns* 8205.
Notes: Rhizome long-creeping. Stipes black to 5.5-6 cm long, naked. Fronds simple, to 32 cm long.

2. **Ecol.**: Common epiphyte in lowland rain forest on small subcanopy trees in rain forest at 70 m altitude.
Specimens: ARFAK PLAINS: Wariori Flood Plain - *Johns* 8220.
Notes: Rhizome short-creeping. Stipes black, naked, to 6cm long. Fronds to 26.5 x 2.9 cm.

3. **Ecol.**: Epiphyte in disturbed lowland rain forest.
Specimens: ARFAK PLAINS: Wariori Flood Plain - *Johns* 8175.
Notes: Margin distinctly lobed. Rhizome short-creeping. Stipe to 65 cm long. Fronds to 32 x 4 cm.

TRIPHLEBIA Baker in Beccari

Triphlebia dimorphophylla Baker in Beccari *Malesia* **3**: 42 (1886). Brause (1920: 144). **Type**: Beccari s.n. Andai, Arfak Mountains. (FI).
= Diplora d'urbillaei

Triphlebia linza (Cesati) Baker in Beccari *Malesia* **3**: 42 (1886).
Asplenium linza Cesati *Rendic. Accad. Sc. Fil. Mat. Napoli* **16**: 26, 29 (1877). **Type**: Beccari s.n. Ramoi. Arfak Mountains. (FI).
= Diplora d'urbillaei

BLECHNACEAE

Copeland, E.B. (1949, 1950). Aspleniaceae and Blechnaceae of New Guinea. *Philip. J. Sci.* **78**: 207-229.

Kato, M. (1992). Taxonomic studies of pteridophytes of Ambon and Seram (Moluccas) Collected by Indonesian-Japanese Botanical Expeditions VIII. Aspleniaceae, Blechnaceae. *J. Fac. Sci. Univ. Tokyo, Sect.* III **15**: 111-133.

BLECHNUM L.

Blechnum borneense C.Chr. *Danks Bot. Ark.* 9(3): 56 (1937). Kato (1992: 145).
Dist.: Borneo to Seram and New Guinea.
Ecol.: Kato (1992) notes that this species is terrestrial in montane forest. Epiphyte in shade in montane forest at 940 m altitude.
Specimens: ARFAK MTS.: Mupi Dessa, trail from Mupi Village to G. Humibou - *Sands* 6859. WITHOUT LOCALITY: *van Royen & Sleumer* 8216 (L).
Notes: Epiphyte with distinct sterile and fertile fronds.

Blechnum finlaysonianum Hooker et Grev. *Icon. Fil.* t. 255 (1831). Wall. *List* n. 2172 (1829). Beccari (1886: 38); Copeland (1950: 224); Holttum (1955: 445); Kato (1992: 145); Lam (1945: 60); Royen (1960: 40, 61). **Type**:
Dist.: Penang, Singapore, Borneo to New Guinea.
Ecol.: Terrestrial in lowland and lower montane forest.
Specimens: RAMOI: *Beccari* s.n.
Notes: Copeland (1950) lists this as a synonym of *Blechnum nudius* Copeland.

Blechnum fraseri (A.Cunn.) Luerss. Flora 292 (1876). Copeland (1950: 227).
Dist.: New Zealand, New Guinea to Philippines.
Specimens: WITHOUT LOCALITY: *van Royen & Sleumer* 8225.
Notes: Kato (1992: 145) lists var. ***philipinense*** Christ from Ceram and New Guinea.

Blechnum orientalis L. *Sp. Pl.* **2**: 1077 (1753). Beccari (1886: 38); Carr. in Seem. (1873: 352); Christ in Warburg (1900: 65); Copeland (1911: 78); Copeland (1950: 224); Copeland (1961: 421); Holttum (1955: 446); Kato (1992: 145); Lauterbach (1929: 11); F.Mueller (1876: 81); Whitmore (1963, 1965: 62).
Dist.: Asia, throughout Malesia to Australia and Polynesia.
Specimens: ARFAK MT.: Putat - *Beccari* s.n. (FI).
Notes: In the first edition of *Sp. Pl.* the names *B. orientale* and *B. occidentale* were accidently transposed. This typographic error was corrected in the second edition.

***Blechnum* spp.**

1. **Specimens**: IRAY II: Widjaja 4472.

2. **Ecol.**: Terrestrial in subalpine forest along mountain path to Mt Koebre at 2500 - 3000 m altitude.
   Specimens: KOEBRE MT.: Path to W of summit - *Johns* 8057.
   Notes: Fronds dimorphic. Fertile fronds: pinnae to 35-40 cm, rachis to 46 cm. Sterile fronds, 24 - 25, rachis to 37 cm.

3. **Ecol.**: Terrestrial in sub-alpine forest.
Specimens: KOEBRE MT.: Along path to west of summit - *Johns* 8081 B.
Notes: Fronds dimorphic. Fertile fronds: stipes black, to 57 cm long; rachis to 57 cm. Sterile fronds: stipes to 23 cm long; rachis to 43 cm, with 38+ pinnae to 5 cm long.

4. **Ecol.** Terrestrial on dry hillside just above swamp to NW of Lake Anggi at 1850 m.
Specimens: ANGGI LAKES: *Johns* 8098.
Notes: Fronds pinnate to 2+ m long. Pinnae sessile to 24 x 1 cm.

STENOCHLAENA J.Smith

Holttum, R.E. (1971). The genus *Stenochlaena* J.Smith with description of a new species. *Amer. Fern J.* **61**: 119-123.

Stenochlaena palustris (Burm. f.) Bedd., *Ferns of Brit. India* 234 (1876). Brass (1941: 287); Bonap. (1921: 232); Copeland (1947: 161); Copeland (1950: 228); Holttum (1932: 235); Kuhn (1889: 5); K. Schumann & Lauterbach (1900: 131); Gepp in Randle (1923: 62); Copeland in C.T. White (1929: 272); Wagner & Grether (1948: 55).
Distr.: Sri Lanka and India trhoughout Indochina to Malesia, N. Australia, Fiji and Polynesia.
Ecol.: In disturbed lowland rain forest.
Specimens: ARFAK PLAINS: ! Km E of Wariori River - *Johns* 8148.

Stenochlaena milnei Underw.
Dist.: Philip[pines and New Guinea to Solomon Islands.
Specimens: ARFAK PLAINS: Wariori River- *Johns* 8172.

CHEIROPLEURIACEAE

CHEIROPLEURIA C.Presl

Cheropleuria bicuspis (Blume) C.Presl. Gepp in Gibbs (1917: 76).
Dist.: From Indo-Chinas, Honshu and Formosa, throughout Malesia to New Guinea.
Ecol.: Terrestrial in high forest, sometimes on rocks in shade. In montane forest.
Specimens: ARFAK PLAINS: On steep ridge between Arfak Plains and G. Itsiwei - *Sands* 6369. ARFAK MTS.: *Gibbs* 6134(BM). WITHOUT LOCALITY: BW 6868..
Notes: The simple-fronded specimens were described as var. *integrifolia*. This variety is not recognised.

CYATHEACEAE

Holttum, R.E. (1963). Cyatheaceae. *Flora Malesiana Series II.* **1**(2): 65-176.
Holttum, R.E. (1981). Addenda, Corrigenda et Emendanda - Cyatheaceae. *Flora Malesiana Series II.* **1**(5): 562-563.

Alsophila arfakensis Gepp = ***Cyathea kanehirae*** Holttum *non* ***Cyathea arfakensis*** Gepp

CYATHEA J.E.Smith

Cyathea angiensis (Gepp) Domin *Acta Bot. Bohem.* **9**: 90 (1930). Holttum (1963: 135); Holttum (1981: 563).

Alsophila angiensis Gepp in Gibbs, *Phytogeog. and Flora Arfak Mts.* 69 (1917). **Type**: L.S. Gibbs 5968. 12/1914. Anggi Lakes. Tree fern to 3 m tall. 2550m. (Holotype BM: K, P, BO).

Dist.: Moluccas (Buru) to New Guinea. Vogelkop, Aitape, Toricelli Mts., Finisterre Mts., E. Highlands, Huon Peninsula, Edie Ck., Goodenough Is.,

Ecol.: Tree fern to 3 m in height. In undergrowth of forest near lakes.

Specimens: ANGGI LAKES: *Gibbs* 5968 (BM).

Notes: Close to *Cyathea contaminans*. Can (?) be distinguished as fronds are in whorls of 4 or 5, and the costae with many densely setiferous scales.

Cyathea archboldii C.Chr. *Brittonia* **2**: 278 (1937).

Cyathea didentata Copel. **Type**: Brass 10971. Lake Habbema. 2750 m. (Isotype: L).

Dist.: Moluccas (Ceram), New Guinea to Bougainville and New Hebrides.

Ecol.: Common along forest edge of *Nothofagus* forest, between 1750 (- 2480) m.

Specimens: NETTOTI RANGE: S slope of Mt Nettoti on Andjai-Wekari path- *van Royen & Sleumer* 7479 (L).

Notes: Fronds to 3 x 1 m. Petiole to 1 m long. 24 pairs of pinnae.

Cyathea arfakensis Gepp in Gibbs, *Phytogeog. and Flora Arfak Mts.* 68 (1917). Holttum (1963: 82). **Type**: *L.S. Gibbs* 6008. 12/1914. Anggi Lakes, on SW ridge. Undergrowth in forest near lake. 2400m. (Holotype BM:).

Hemitelia arfakensis Alderw. *Bull. Jard. Bot. Btzg.* II. n. 28: 26 (1918).

Dist.: Endemic.

Ecol.: Undergrowth in forest.

Specimens: ANGGI LAKES: *Gibbs* 6008 (BM); *Kanehira & Hatusima* 13596.

Cyathea biformis (Rosenst.) Copel. *Philip. J. Sci.* Bot. **2**: 123 (1941). Brass (1941: 282, 289, 290, 298); Cretz (1941: 300); Copeland (1948: 117); Domin (1930: 48).

Alsophila biformis Rosenst. *Rep. Sp. Nov.* **9**: 423 (1911). Brause (1920: 74). **Type**: King 57.

Gymnosphaera biformis (Rosenst.) Copel. (1917).

Polybotrya arfakensis Gepp. in Gibbs Arfak (1917: 71). **Type**: L.S. Gibbs 5984. Anggi Lakes. 2100 m. (BM).

Dist.: NW to SE New Guinea.

Ecol.: Climbing in forest. Clasping trunk.

Specimens: ANGGI LAKES: *Gibbs* 5984 (BM).

Cyathea contaminans (Wall. ex Hook.) Copel. *Philip. J. Sc. Bot.* **4**: 60 (1909). Holttum (1963: 135).

Polyodium contaminans Wall. *Cat.* (1828). nomen. **Type**: *Wallich* 320. Penang (K).

Dist.: Burma, throughout Malesia.

Ecol.: Common to abundant in clearings and open places in forest. 200 to 1600 m.

Specimens: WITHOUT LOCALITY: *Kanehira & Hatusima* 11871.

Notes: *Cyathea angiensis* should, perhaps, be united with this species. Young specimens are often quite scaly. There are several varieties but these probably only reflect environmental effects.

Cyathea sp. aff. ***Cyathea contaminans*** (Wall. ex Hook.) Copel.

Ecol.: On flat ground along valley floor in lower montane forest. 810 m alt.

Specimens: ARFAK MTS.: Mupi Dessa Trail - *Sands* 6831.

Vernacular: 'Selowa' in Hatam Language.

Notes: Trunks to 17 m. Fronds to 5 m long with 15-18 pairs of pinnae. Has rigid dark branching hairs/scales at base of pinnae.

Cyathea costalisora Copel. *Univ. Cal. Publ. Bot.* **18**: 218 (1942). Copel. *Philip. J. Sc.* **77**: 101. pl. 1 (1947). Holttum (1963: 87). **Type**: *L.J. Brass* 9488. Lake Habbema. Irian Jaya. (A: BO, BM, L).

Dist.: Endemic to Irian Jaya. Lake Habbema and Anggi Lakes (1900 m).

Ecol.: Near lake in forest. Anggi Lakes - 1900 m and Lake Habbema - 3225 m .

Specimens: ANGGI LAKES: *Kanehira & Hatusima* 13762.

Notes: The specimen from Anggi Lakes differs from the type collection which has very fine brown scales at the base of the pinnae.

Cyathea everta Copel. *Univ. Cal. Publ. Bot.* **18**: 218 (1942). Copel. *Philip. J. Sc.* **77**: 103. pl. 2 (1947). Holttum (1963: 85). **Type**: *L.J. Brass* 10712. Lake Habbema. Irian Jaya. (Holotype A: BO, L, UC).

Cyathea globosora Copel. *Univ. Cal. Publ. Bot.* **18**: 218 (1942). Copel. *Philip. J. Sc.* **77**: 106. pl. 4 (1947).

Dist.: Irian Jaya, Kaugel Valley, Mt Piora, .

Ecol.: From 1400-2800 m in oak forest, on clay, and mossy forest (mixed conifer). Rather scarce.

Specimens: ANGGI LAKES: *Gibbs* 5932 (BM). Mt Sarumto - Sleumer & Vink BW 14342 (L).

Notes: This species is closely related to *Cyathea rigens.* The stipe is light brown with numerous small spines (not sharp) along it. Basal scales on stipe are very dark brown to black fringed with a light brown region.

Cyathea felina (Roxb.) C.V. Morton *Contrib. U.S. Nat. Herb.* **38**(7): 344 (19). **Type**: Roxburgh. Amboina. (**Lectotype** G).

Cyathea sangirensis (Christ) Copel.

Dist.: Moluccas, throughout New Guinea - Aru Is., mainland New Guinea to Louisiade Arch.

Ecol.: A large lowland tree fern often in secondary habitats. Or on steep sided valley in lowland rain forest. Slope muddy. 250 m alt.

Specimens: ANDAI: Forest Reserve - *Sands* 6174. WITHOUT LOCALITY: *Kanehira & Hatusima* 13286.

Notes: Trunks to 2 m. Fronds to 4 m long, stipes with dark reddish brown scales on one side and other side with soft buff scales.

Cyathea fusca Baker in Beccari, Malesia 3: 31 (1886). Alderw. (1908: 19); Alderw. (1917: 25); Copel. (1947: 100); Gibbs (1917: 69); Holttum (1963: 154). **Type**: *d'Albertis s.n.* Fly River. Papua. (K).

Dist.: Endemic.

Ecol.: Undergrowth of forest near river.

Specimens: ANGGI LAKES: Gibbs 5932 (BM).

Notes: This species is closely related to *C. werneri* Rosenst. Holttum records this species as restricted to Papua New Guinea with no reference to Gibbs 5932, the records published in 1917.

Cyathea gibbsiae Copeland *Philip. Journ. Sci.* **38**: 129 (1929). Brause (1917: 117); Cretz (1941: 305); Domin (1930: 118).

Polybotrya arfakensis Gepp in Gibbs *Contr. Flora Arfak Mts.* 71 (1917). **Type**: Gibbs 5984. (FI).

Specimens: ARFAK MTS.: Gibbs 5984.

Notes: Copeland (1948) lists this as a possible synonym of ***Cyathea biformis.***

Cyathea ***intermedia*** (Mett.) Copeland *Univ. Calif. Publ. Bot.* **16**: 357 (1929). Copeland (1948: 114); Cretz (1941: 307); Domin (1930: 126).
Alsophila intermedia Mett. *Ann. Sc. Nat.* **15**: 83 (1891). Beccari (1886: 31); Beccari in d'Albertis (1880); Brause (1920: 71); Cesati (1877: 24).
Dist.: New Guinea to New Caledonia.
Specimens: RAMOI: *Beccari s.n.* (FI).

Cyathea ***kanehirae*** Holttum *Flora Malesiana* ser. II, **1**(2): 113 (1963).
Alsophila arfakensis Gepp in Gibbs, *Phytogeog. and Flora Arfak Mts.* 70 (1917). **Type**: L.S. Gibbs 5990. Dec. 1914. Anggi Lakes. Arfak Mts. Common in undergrowth of moss forest. 2100 - 2550 m. (Holotype BM: K, P).
Dist.: Endemic, Irian Jaya.
Ecol.: Common in undergrowth of moss forest. 1600-2700 m.
Specimens: ANGGI LAKES: *Gibbs* 5990 (BM); *Kanehira & Hatusima* 13499 (BO, A).
Notes: A slender tree fern to 1 cm diameter and 1 m tall.

Cyathea ***ledermannii*** Brause *Bot. Jahrb.* **56**: 56 (1920). Copel. *Philip. J. Sc.* **77**: 119 (1947); Holttum (1963: 83). **Type**: Ledermann 9651. Sepik Region. Papua New Guinea. (B).
Dist.: New Guinea endemic.
Ecol.: 200-1000 m in rain forest and mossy forest. Locally abundant.
Specimens: WITHOUT LOCALITY: *van Royen & Sleumer* 8222.

Cyathea magna Copel. *Univ. Cal. Publ. Bot.* **18**: 218 (1942). Copel. *Philip. J. Sc.* **77**: 110. pl. 7 (1947); Holttum (1963: 85). **Type**: *L.J. Brass* 10278. Lake Habbema. Irian Jaya. 2250 m. (Holotype A: BO, K, L, BM, UC, MICH).
Alsophila tomentosa var. ***novoguineensis*** Rosenst. *Fedde Rep.* **5**: 34 (1908).
Cyathea ordinata Copel. *Philip. J. Sc.* **77**: 109 (1947).
Dist.: Endemic in New Guinea.
Ecol.: In open areas in fagaceous forest and in grassland. 1700-2750 m. Locally common.
Specimens: WITHOUT LOCALITY: *Kanehira & Hatusima* 13763.
Notes: This species is closely related to *C. tomentosa* from Java.

Cyathea pulcherrima Copel. *Univ. Cal. Publ. Bot.* **18**: 219 (1942). Copel. *Philip. J. Sc.* **77**: 110 (1947); Holttum (1963: 126). **Type**: *L.J. Brass* 8940. Cycloop Mts, Irian Jaya. 575 m. (Holotype UC: K, L, BM).
Dist.: Endemic. Throughout New Guinea to Admiralty Island.
Ecol.: In lowland and lower monatne forest. From 100-1100 m altitude.
Specimens: ARFAK PLAINS: In forest S of Plains - Sands 6405. WITHOUT LOCALITY: *Kanehira & Hatusima* 12173.

Cyathea tripinnata Copel. Philip. J. Sci. 1. Suppl. 4, Bot. 251 (1906). Holttum (1963: 140).
Dist.: Java, Malaya, Borneo, Philippines, Moluccas, Irian Jaya.
Ecol.: In forest, 250-1700 m.
Specimens: *Kanehira & Hatusima* 12622; *Kanehira & Hatusima* 13242.
Notes: Holttum (1963: 140) does not record this species from New Guinea.

Cyathea werneri Rosenst.
Ecol.: Terrestrial on lower valley sides in mature rain forest on uplifted coral at 300 m altitude.

Specimens: ARFAK PLAINS: Valley to S - Johns 8198. ANDAI FOREST: 2.5 km west of Andai - *Johns* 8108. ANDAI: 2.5 km W of Andai - *Johns* 8114.
Notes: Trunks to 2-3 m tall and and 6 cm diameter. Fronds with short stipe to 10-15(-20) cm long, densely covered with scales. Rachis to 1.80 - 2.5 m long. Lower pinnae very short to 5 cm long. Very small fronds near base. Pinnae length increasing to 12th pinna. See sketch of pinna.

Previously only known from S and E New Guinea.

Cyathea **sp. nov. 1**
Ecol.: On ridge top in partial shade in lower montane forest. 675 m alt.
Specimens: ARFAK PLAINS: Steep ridges to G. Itsiwei - *Sands* 6424.
Notes: Trunks to 10 m. Fronds to 1.6 m long with 16-18 pairs of pinnae.

Subsection *Fourniera*

This species is related to ***C. celebica*** Blume, ***C. tripinnata*** Copel. and ***C. macrophylla*** Domin.

Cyathea **sp. nov. 2**
Ecol.: In forest at lower level of montane species. 80-150 m altitude.
Specimens: ARFAK PLAINS: Wariori River - *Johns* 8193.
Notes: Tree fern to 8-10 m tall. Stipe to 24-30 cm long, grooved above, with a line of 3-4 prominent, white aerophores, brown, base spiny. Rachis to 383 cm long with 18 - 21 pairs pinnae. Longest pinna to 109 cm. Apical 4-5 pinnae not fertile. See table with specimens.

Subsection *Fourniera*

This species is related to ***C. celebica*** Blume, ***C. tripinnata*** Copel. and ***C. macrophylla*** Domin.

Cyathea **sp. nov. 3**
Ecol.: Terrestrial on limestone to N of Kebar at 600 m.
Specimens: KEBAR VALLEY: *Johns* 8349b.
Notes: Stipes 35-58 cm long, densely covered with spines. Rachis to 250-280 cm long, lower pinnae widely spaced. Pinnae alternate.

This species is **'quite unlike'** any other collections of *Cyathea* from New Guinea.

Cyathea **sp. nov. 4**
Ecol.: Above Lake in grassland and scrub at 1950 m altitude.
Specimens: KOEBRE MT.: Mountain path from Lake Anggi - *Johns* 8092.
Notes: Trunk to 1 m tall. Stipe scaly to 34-60 cm, very spiny. Rachis reddish, spiny, with 10-20 pinnae.

Subsection *Fourniera*

This species is related to ***C. celebica*** Blume, ***C. tripinnata*** Copel. and ***C. macrophylla*** Domin.

Cyathea **sp. nov. 5**
Ecol.: Terrestrial in ridge top forest on limestone escarpment at 900 m altitude.
Specimens: KEBAR VALLEY: On ridge above Apni Valley east of Andjai 2 - *Johns* 8285.
Notes: Small trunk to 15 cm tall and 4.5 cm diameter. Fronds dimorphic. Sterile fronds: stipes to 35 cm long; 8 sub-opposite pairs.

Cyathea **spp.**

1. **Ecol.**: Terrestrial in small remnant of rain forest on very wet garden area NW of Lake.
   Specimens: ANGGI LAKES: *Johns* 8100.

Notes: Trunk to 2 m tall with 5-6 fronds. Base of stipe and apex of trunk covered with very broad grey scales. Fronds to 2+ m long. Rare. Only 2 plants seen.

2. **Ecol.**: Terrestrial in disturbed lowland tropical rain forest at c. 50 m altitude.
Specimens: ARFAK PLAINS: *Johns* 8128.
Notes: Trunk to 2-3 m tall. Rachis to 160 cm. Stipe short to 2 cm. Plant with 11 to 12 pinnae. See table with collection.

3. **Ecol.**: Terrestrial in lowland rain forest with trunk to 7 m tall and 25 cm diam. On Wariori Flood Plain at 70 m altitude.
Specimens: ARFAK PLAINS: Wariori Flood Plain - *Johns* 8217.
Notes: Base of stipe densely scaly. Stipe to 45-53 cm long. Rachis to 153 cm long with 13-15 pinnae, largest to 42-53 x 15-18 cm. See table for data.

4. **Ecol.**: Terrestrial tree fern in disturbed forest on limestone outcrop at 600 m altitude.
Specimens: KEBAR VALLEY: Anjai 2, on limestone outcrop - *Johns* 8362.
Notes: Trunk to 2-3 m tall. Stipes from 55-82 cm long. Rachis to 139-177 cm long with 9-10 pinnae and a single terminal pinna. Margins serrate at vein ends. See table with collection.

5. **Ecol.**: Terrestrial in ridge top forest along limestone escarpment at 900 m altitude.
Specimens: KEBAR VALLEY: E of Anjai 2, Apni Valley - *Johns* 8266.
Notes: Trunk to 1 m tall with sticky sap. Stipe bases with long black and dense short spines. Stipe to 63 cm long, bases densely scaly, groved above with upper surface lacking spines. Fronds bipinnate, to 1.5 m long. Rachis black.

6. **Ecol.**: In ridge forest from Warmare to Mokwam at 800-850 m altitude.
Specimens: ARFAK MTS.: Mokwam Trail - *Johns* 7723.
Notes: Trunk to 1-2 m tall. Fronds bipinnate-pinnatifid. Stipe short with reduced basal pinnae. Lower pinnae to 75 cm long. Largest pinna to 85 cm long.

DAVALLIACEAE

Davalliaceae spp.

1. **Ecol.**: In wet patches in forest on uplifted coral at 250 m altitude.
Specimens: GUNUNG MEJA: *Johns* 7666.
Notes: Fronds to 1.5 m tall with 14 pinnae on each side of rachis. Lowest pinnae to 19 cm long, next pinna 30 cm long. Sori with small brown cup-shaped indusium.

2. **Ecol.**: Terrestrial along small valley at 300 m altitude.
Specimens: ANDAI FOREST: In mature rain forest on uplifted coral - *Johns* 8121.
Notes: Plants to 2-3 m tall. Stipe, spiny, long, grooved, to 1.5 m long. Bipinnate/pinnatifid. Pinnae widely spaced with 12-15 pinnae.

3. **Ecol.**: Epiphyte in forest at 1700 - 1940 m altitude.
Specimen: ANGGI LAKES: Iray - *Johns* 7997.
Notes: Epiphyte, rhizome densely covered with scales. Stipe black to 10 cm long. Fronds bipinnatifid, 16 x 13 cms.

DAVALLIA J.E.Smith

Davallia dissecta J.Sm. Gepp in Gibbs (1917: 72); Rosenst. (1913: 524).
Dist.: Java, Sumatra to New Guinea.
Ecol.: Epiphyte in forest patches. 2100 m.
Specimens: ANGGI LAKES: *Gibbs* 5592 (BM).

***Davallia* spp.**

1. **Ecol.**: Epiphyte on tree trunks or terrestrial on moss cushions in subalpine forest along track to Mt Koebre at 2500-3000 m altitude.
   Specimens: MT KOEBRE: Path to W of summit - *Johns* 8056.
   Notes: Fronds dimorphic, reduced in fertile fronds. Sori terminal on lobes with small white translucent scales.

2. **Ecol.**: Epiphyte on tree trunks in ridge top lower montane forest dominated by *Araucaria*, *Lithocarpus* and *Litsea* at 650 m altitude.
   Specimens: ARFAK MTS.: Mokwam Trail - *Johns* 7781.
   Notes: Fronds dimorphic, fertile frondds finely divided, sterilefronds broader. Rachis to 5-6 cm long. 10-12 cm long.

HUMATA Cav.

Humata alpina (Blume) Moore *Ind. Fil.* Gepp. in Gibbs (1917: 72); Rosenst. (1908: 370); K. Schumann & Lauterbach (1905: 40); Wagner & Grether (1917: 41).
Dist.: Mascarines and India to Malesia and Polynesia.
Ecol.: Epiphyte in forest patch. 2100 m.
Specimens: ANGGI LAKES: *Gibbs* 5674 (BM).

Humata neoguinensis C.Chr. *Ind. Fil.* 354 (1908). Brause (1920: 122); Copeland (1940: 353); Gepp in Gibbs (1917: 72). **Type**:
Dist.: New Guinea. Arfak to Hellwig Mts.
Ecol.: Epiphytic in forest or creeping long forest edge. 2100 m.
Specimens: ANGGI LAKES: *Gibbs* 5588 (BM).

Humata pusilla (Mett.) Carr. in Seem. *Fl. Vit.* 335 (1873). Copeland (1940: 351); Gepp. in Gibbs (1917: 72); Schlechter in Brause (1912: 4); K. Schumann & Lauterbach (1905: 354); Wagner & Grether (1948: 41).
Davallia pusilla Mett. (1861).
Dist.: New Guinea to Melanesia.
Ecol.: Epiphyte in forest. Common on tree bases in lowland rain forest and in *Castanopsis* - *Lithocarpus* forest. To 600 m altitude.
Specimens: ANGGI LAKES: *Gibbs* 5960 (BM).

***Humata* sp.**

1. **Ecol.** Epiphyte in shade in montane forest at 900 m altitude.
   Specimens: ARFAK MTS.: Mupi Dessa Trail - *Sands* 6827.
   Notes: Epiphytic, rhizome creeping, fronds coriaceous.

DENNSTAEDTIACEAE

Copeland, E.B. (1949). Pteridaceae of New Guinea. *Philip. J. Sci.* **78**: 5-40. In New Guinea eight genera with c. 40 species.

Dennstaedtiaceae indet.

1. **Ecol.**: Terrestrial in small patch along minor valley on uplifted coral in mature rain forest at 300 m altitude.
   Specimens: ANDAI: 2.5 km W of Andai - *Johns* 8111.
   Notes: Rhizome short-creeping, densely covered with long, brown scales. Stipe to 50-60 cm, deeply grooved. Adaxial side of lower pinnae longer. Pinnae broadly serrate. Sori brownish.

2. **Ecol.**: In montane forest near Iray at 1800-1940 m.
   Specimens: ANGGI: Iray - *Johns* 7910 A.
   Notes: Rhizome long-creeping, covered with brown multicellular hairs. Stipe deeply grooved to 71 cm. Fronds tripinnate with rachis densely hairy.

DENNSTAEDTIA Bernhardi

A genus with c. 12 spp. in New Guinea.

Dennstaedtia glabrata (Ces.) C.Chr. *Ind. Fil.* 217 (1908). Alderw. (1909: 145); Brause (1950: 12); Copeland (1950: 12); Holttum (1963: 162).
Dicksonia glabrata Ces. *Rend. Ac. Napoli* **16**: 24, 28 (1877).
*? **Dennstaedtia tripinnatifida*** Copel. Type: Brass 3043. San Cristobal. Solomon Is. (L).
Dist.: Malay Peninsula to Fiji. Widespread throughout New Guinea. Pioneer Bivouac, Doormann R., Vanimo, Huon Peninsula, Ekuti Divide, Mt Dayman., Fly R., Rossel, Normanby,
Ecol.: Common in undergrowth of rainforest and transiyional forests to *Castanopsis* forest. (200-)450-840(-2200) m altitude.
Specimens: ARFAK MTS.: Putat - *Beccari s.n.* (FI).
Notes: Stipes to 1 m long+ fronds to 2 m. Fronds 3 pinnate-pinnatifid. Base of stipe with scattered spines.

Dennstaedtia scabra (Wall. ex Hook.) Moore. Alderw. (1909: 143).
Dicksonia scabra Wall. ex Hook. *Sp. Fil.* **1**: 80 (1844).
Dist.: Sri Lanka, India and China to New Guinea.
Specimens: ANDAI: *Beccari s.n.* (FI).
Notes: Also *D. scabra* Alderw. (1909: 143).

HISTIOPTERIS (Agardh) J. Smith

Copeland (1950: 29-30) includes a key to the two species recorded from New Guinea.

Histiopteris incisa (Thunb.) J.Sm. *Hist. Fil.* 295 (1875). Copeland (1950: 30); Gepp. in Gibbs (1917: 73); K.Schumann & Lauterbach (1900: 135); Wagner & Grether (1948: 37).
Pteris incisa Thunb. **Type**:
Dist.: Pantropical and subtropical. Tropical Africa to Japan. Throughout Malesia to New Zealand. Widespread in New Guinea.

Ecol.: Terrestrial in secondary forest and old garden areas. (950-)1410-1940 m. In secondary sites at lower alttiudes - on volcanis debris in PNG.
Specimens: ANGGI LAKES: *Gibbs* 5669 (BM); Between end of Airstrip and Lake Giggi - *Johns* 7914.
Notes: Stipe to 52 cm long, dense black. Fronds bipinnate-pinnatifid. Sorus continuos along margins of pinnae except near axis.

HYPOLEPIS Bernhardi

Brownsey, P.J. (1987). A review of the fern genus *Hypolepis* (Dennstaedtiaceae) in the Malesian and Pacific regions. *Blumea* **32**(2): 227-276.

Hypolepis cf. ***Hypolepis scabristipes*** Brownsey *Blumea* **32**(2): 238 (1987). **Type**: Millar & Holttum NGF 18800. Wau. Papua New Guinea. 1050 m. (L: BRI, CANB, K).
Dist.: Malaya, Sumatra to Papua New Guinea.
Ecol.: Srambling or climbing plant. In montane forest near Iray at 1800-1940 m.
Specimens: ANGGI: Iray - *Johns* 7910 B.
Notes: Rhizome long-creeping, covered with brown multicellular hairs. Stipe deeply grooved to 71 cm. Fronds tripinnate with rachis densely hairy.
First record from Irian Jaya.

MICROLEPIA Presl.

Copeland (1950) includes detailed notes and keys to the species he recognised from New Guinea. Six species in New Guinea.

Microlepia speluncae (L.) T. Moore *Ind. Fil.* 43 (1857). Carr. in Seem. (1873: 340); Christ in Warburg (1900: 87); Copeland (1950: 14); Lauterbach (1930: 431); K. Schumann & Lauterbach (1900: 122); Sledge (1957: 524, 527); Wagner & Grether (1948: 33).
Polypodium speluncae L. *Sp. Pl.* 1093 (1953). **Type**:
Dist.: Widespread in Old World Tropics.
Ecol.: Terrestrial on limestone and on banks in swamp forest. From sea-level to 60 m. Secondary at higher altitudes: 750-950 m.
Specimens: *Sands* 6159.
Notes: Rhizome short-creeping. Fronds to 2 m long with stipe to 1 m.
Young fronds eaten in West New Britain.
Sledge (1956) made a detailed study of this species and recognised three varieties.

PAESIA St. Hilaire

Copeland (1950: 28-29) inlcudes notes on two species from New Guinea. Holttum (1959: 454) discusses the genus.

Paesia radula (Baker) C.Chr. *Ind. Fil.* Brause (1920: 163); Copeland (1950: 29); Gepp in Gibbs (1917: 74); Holttum (1959: 454); Posth. in Steenis (1934: 172).
Orthiopteris scaberula Rosenstock (1913: 142).
Dist.: Sumatra, Borneo and Sulawesi to New Guinea. In New Guinea widespread.

Ecol.: Forming clumps near forest margins up to 2800 m altitude. Forming clumps to 1.5 m diameter.
Specimens: ANGGI LAKES: *Gibbs* 5956 (BM).

PTERIDIUM Scopoli

Pteridium aquilinum (L.) Kuhn in Decken *Reisen. Ost. Afr.Bot. 3* **3**: 11 (1879). Brause (1920: 163); Copeland (1911: 85); Copeland (1950: 29); Lauterbach (1930: 465); Wagner & Grether (1948: 37).
Dist.: Cosmopolitan.
Ecol.: Very common in grassland andsecondary habitats. In New Guinea on Limestone at 275 m alt.
Specimens: ANGGI LAKES: BW 623. KOEBRE: *Gibbs* 5599 (BM). *van Royen & Sleumer* 5091.

Pteridium aquilinum (L.) Kuhn var. ***wightianum*** (J. Agardh) R.M. Tryon
Ecol.: Terrestrial along lake margins, at 1700-1940 m altitude. Also along water-logged lake margin strewn with boulders.
Specimens: ANGGI LAKES: Lake Giggi - *Johns* 7972. Between Airstrip at Iray and Lake Giggi - *Johns* 7913.
Notes: Stipes to 75 cm. Rachis to 80 cm with 9 pairs of pinnae and terminal pinna.
Rhizomes long-creeping. Stipes with dense, branched, hairs at base.
Very hairy abaxially.

Pteridium **spp.**

1. **Ecol.**: In disturbed areas near Lake Anggi Gigi.
Specimens: ANGGI LAKES: Iray - *Johns* 7909.
Notes: Stipe to 105 cm. Base of stipe covered with long, divided scales.

2. **Ecol.**: In montane forest near Iray at 1800-1940 m.
Specimens: ANGGI: Iray - *Johns* 7910 A.
Notes: Rhizome long-creeping, covered with brown multicellular hairs. Stipe deeply grooved to 71 cm. Fronds tripinnate with rachis densely hairy.

ORTHIOPTERIS Copeland (SACCOLOMA Kaulf.)

Orthiopteris campylura (Kunze) Copeland
Saccoloma minus (Hooker) C.Chr.
Orthiopteris kingii (Tryon 1962).
Dist.: Vogelkop to Manus (Mt Dremsel).
Ecol.: Terrestrial on damp slopes in hill forest c. 600 m altitude.
Specimens: WITHOUT LOCALITY: *Beccari* 12347. (FI).

DICKSONIACEAE

DICKSONIA L'Herit.

Dicksonia archboldii Copel. *Philip. J. Sc.* **78**: 6. pl.1 (1949); Holttum (1963: 160). **Type**: *L.J. Brass* 10970. Lake Habbema, Irian Jaya. 2750 m. (Holotype MICH: L).

Dist.: Rare in Irian Jaya. Known only from type Lake Habbema and Anggi Lakes. The Anggi Lakes collection is smaller than the Type collection.
Ecol.: Locally abundant in forest and along locally common along forest margins. 2400-2750 m.
Specimens: ANGGI LAKES: Mt. Sensenemes - *Sleumer & Vink* BW 14193. (L).
Notes: The hairs on the costae are red and ascending. Apex of stem with very dense, fine hairs.

Dicksonia sciurus C.Chr. *Brittonia* **2**: 283 (1937). Copel. *Philip. J. Sc.* **78**: 6 (1949); Holttum (1963: 160). **Type**: L.J. Brass 4991. Mt Tafa. Papua New Guinea. (Holotype BM: BO, MICH, BRI, UC).
Dist.: Waigeo Is, throughout New Guinea, several collections mainly from E to Solomon Is.
Ecol.: Common in ridge forest. 1800-2400 m.
Specimens: Croxall 9.

***Dicksonia* sp.**
1. **Ecol.**: In subalpine rainforest and shrubbery along track to Mt Koebre at 2500-3000 m altitude. Uncommon to rare.
Specimens: MT. KOEBRE: *Johns* 8050.
Notes: Small trunk to 2 m tall. Fronds with 24 pinnae, the lowest 5-6 reduced to 5-8 cm long. Longest to 36 x 14 cm. Costa very hairy, fronds very brittle. Sori marginal protcetde by small cup-shaped indusia.

CULCITA Presl

Culcita straminea (Labill.) Maxon, *J. Wash. Acad. Sc.* **12**: 457 (1922). Copel. (1958: 56); Holttum (1963: 169); Wagner & Grether (1948: 33).
Dicksonia straminea Labill. *Sert. Austr. Cal.* 7, t.10 (1824). **Type**: Labillardiere. New Caledonia. (P).
Dist.: Philippines, New Guinea to Polynesia.
Ecol.: In lowland areas, in secondary forest and on river banks. Also terrestrial in ridge top forest on limestone escarpment at 900 m altitude.
Specimens: KEBAR VALLEY: On ridge above Apni Valley east of Andjai 2 - *Johns* 8255.
Notes: Rhizome forming a short trunk to 20 cm long. Trunk densely covered with golden hairs. Fronds to 2-2.5 m long, pinnae alternate. Sparingly fertile on basiscopic side of costa.

CYSTODIUM J. Smith

Cystodium sorbifolium (Sm.) J.Sm. in Hook. *Gen. Fil.* t. 96 (1841). Copeland (1947: 49); Copeland (1950: 7); Holttum (1963: 165); van Royen (1960: 40,61); Wagner & Grether (1948: 33).
Dicksonia sorbifolia Sm. in Rees, *Cyclop* **11**: n.p. (1808). **Type**: C. Smith Honimoa, Ceram. (K).
Dist.: Borneo, Sulawesi, New Guinea to Manus and Louisiade Archipelago.
Ecol.: In lowland forest, often near rivers, to 400 m.
Specimens: RAMOI: *Beccari* s.n. (FI).

DIPTERIDACEAE

Johns, R.J. and Bellamy, A. (1979). The ferns and fern allies of Papua New Guinea. Part 5. The Dipteridaceae. Pp. 19.1-19.10. Office of Forests. P.N.G.

DIPTERIS Reinw.

Dipteris conjugata Reinw. *Syll. Pl. Nov. Ratisb.* **2**: 3 (1825/1826). Carr. in Seem. (1873: 369); Copeland (1947: 177); Gepp in Gibbs (1917: 72); Gepp in Rendle (1923: 52); Lam (1945: 300); Royen (1960: 36,61); Seward & Dale (1924: 15).

Dist.: Asia to Polynesia. Throughout Irian Jaya - Sorong, Idenburg River, Doorman-top, Cycloop Mts.

Ecol.: On banks and steep gravel slopes. From (875-) 2100 - 2550 m altitude.

Specimens: ARFAK PLAINS: Mountains to South - Sands 6370. ARFAK MTS.: Andai - *Beccari s.n.* (FI). ANGGI LAKES: *Gibbs* 5913 (BM). KEBAR VALLEY: Kalkman BW 6379; Koster BW 8038. WITHOUT LOCALITY: BW 10318. Teysmann s.n. (BO).

Notes: Sands 6370 had an exceptionally long stipe to 4 m.

Dipteris novoguineensis Posthumus *Rec. Trav. Bot. Neerl.* **259**: 248. f.1 (1928). Brass (1941: 335); Lam (1945: 137); Lam & Verheys ((1945: 558). **Type**: Lam. DoormanTop. Irian Jaya.
Dipteris papilioniformis Kjellberg. **Type**: Rubbing of specimen from B. Poka, Pindjang. 2000 - 3000 m (BO).

Dist.: Throughout New Guinea.

Ecol.: Common, particularly in disturbed areas of montane forest to 2000 m altitude.

Specimens: MT. NETTOTI: *van Royen & Sleumer* 7937 (BO).

DRYOPTERIDACEAE

CYCLOPELTIS J. Smith

Cyclopeltis novoguineensis Rosenst. *Fedde Repert. Spec. Nov. Regni Vege.* **10**: 329 (1912). Alderw. (1917: 139); Holttum (1991: 118). **Type**: Bamler 46. Logaueng. (BM).
Aspidium semicordatum var. ***biauriculatum*** Baker *J. Linn. Soc. Bot.* **19**: 294 (1882).
Cyclopeltis presliana var. ***biauriculata*** (Baker) Alderw. (1908: 18). **Type**: Commins s.n. Solomon Is. (K).

Dist.: East New Guinea, Bismark Arch. to Solomon Islands. First record from Irian Jaya.

Ecol.: Terrestrial on steep limestone in forest, and on uplifted coral in secondary forest.

Specimens: ARFAK MTS.: Andai - *Johns* 8387. GUNUNG MEJA: *Johns* 7659.

? = *Cyclopeltis novoguineensis* Rosenst.

Specimens: ANDAI: Beccari s.n.

Notes: This collection was identified as ***Aspidium semicordatum*** Swartz *Schrad. Journ.* **1800**(2): 31 (1801). This species is from the New World.

DRYOPTERIS Adans.

Dryopteris leucolepis
Specimens: WITHOUT LOCALITY: *Beccari* 12660.

Dryopteris phaeostigma (Cesati) C.Chr. *Index*. Brause (1920:).
Aspidium phaeostigma
Specimens: ARFAK MTS.: Putat - *Beccari s.n.* WITHOUT LOCALITY: *Beccari* 12534.

Dryopteris polymorphum
Specimens: ARFAK MTS.: Andai - *Beccari s.n.* (FI).

Dryopteris sparsa (Don) Kuntze
Ecol.: From 1700 - 1940 m altitude.
Specimens: IRAY: *Johns* 7991.
Notes: Stipe to 40 cm long with very large brown scales. Largest pinnae to 21 x 5-6 cm. First record from Irian Jaya.

Dryopteris tuberculata (Cesati) C.Chr.
Nephrodium tuberculatum Cesati. **Type**: *Beccari s.n.* Andai, Arfak Mts. (FI, K).
Specimens: ARFAK MTS.: Andai - *Beccari s.n.* (FI, K). *Beccari* 12539 (FI).

***Dryopteris* spp.**

1. **Specimens**: IRAY: *Widjaja* 4210 (BO).

2. **Specimens**: WITHOUT LOCALITY: *van Royen & Sleumer* 6783.

3. **Ecol.**: Terrestrial in Nothofagus forest at 1800 m on S slopes of Mt Nettoti (decent to Andjai).
Specimens: KEBAR VALLEY: Mt Nettoti - *van Royen & Sleumer* 8225A (L).

4. **Ecol.**: A common but scattered terrestrial in primary forest at 1500 m altitude.
Specimens: ARFAK MTS.: Mt Nerimbau - *Ch. Koster* BW 13860 (L).
Notes: Hattam Language - Hange.

PLEOCNEMIA Presl

Pleocnemia andainensis (Baker) Holttum *Kew Bull.* **29**: 350 (1974). Holttum (1991: 15).
Polypodium (*Dictyopteris*) ***andaiense*** Baker in Beccari, *Malesia* **3**: 45 (1886). **Type**: Beccari s.n. Andai, Irian Jaya. (FI: K).
Dictyopteris anadaiensis (Baker) Alderw. (1908: 514).
Dryopteris leuzeana (Gaud.) Kuntze s.l.
Dist.: Endemic in VogelKop Peninsula.
Ecol.: Not known.
Specimens: ARFAK MTS.: Andai - Beccari s.n. (FI).
Notes: Only known from 2 collections. Type and van Royen 3144 from Sorong.

Pleocnemia sp. nr. Pleocnemia andainensis
Ecol.: Terrestrial along small valley in mature rain forest on uplifted coral at 300 m altitude.
Specimens: ANDAI FOREST: *Johns* 8104.

Notes: Rhizome forming a small trunk.

Similar to ***P. andainensis*** but with smaller pinnules/lobe. That species is only known from two incomplete collections from Irian Jaya.

Pleocnemia irregularis (Presl) Holttum *Kew Bull.* **29**: 347 (1974). Holttum (1991: 12).

Arcypteris irregularis (Presl) Ching.

Dist.: S, Burma, Indochina, throughout Malesia to Solomon Islands, Carolines and Fiji.

Ecol.: In lowland rain forest on steep valley slopes, often in seasonally dry regions.

Specimens: GUNONG MEJA: Widjaja 3140.

***Pleocnemia* sp.**

ARFAK PLAINS: Wariori Flood Plain - *Johns* 8154.

TECTARIA GROUP (Dryopteridaceae)

Holttum, R.E. (199). *Tectaria* Group. *Flora Malesiana* Ser. II. Vol. **2**:

CYCLOPELTIS

Cyclopeltis* aff. *Cyclopeltis cumingiana

Dist.:

Ecol.: In secondary forest on uplifted coral along road side at 110 m altitude.

Specimens: **GUNUNG MEJA**: Johns 7659.

Notes: Small stem to 6 cm high. Fronds dimorphic. Fertile fronds to 40 x 2 cm. Sterile fronds to 30 x 4-5 cm.

PTERIDRYS C.Chr. & Ching

Pteridrys microtheca (Fee) C. Chr. & Ching

Ecol.: Near fast flowing stream at 570 m.

Specimens: ARFAK MTS. : Mupi Valley System, along track Mupi to Humelbou - *Mayar* 379.

Notes: Fronds to 130 cm long.

TECTARIA Cavanilles

Tectaria angulata

Aspidium angulatum (Willd.) Mett.

Polypodium angulatum Willd. **Type**: Ventenat s.n. Java (B. Herb Willd. 19641).

Specimens: ANDAI: *Beccari s.n.* (FI).

Tectaria aurita

Specimens: *van Royen & Sleumer* 3992.

Tectaria beccariana (Ces.) C. Chr. *Ind. Fil.* **3**: 177 (1934). Holttum (1991: 93).

Aspidium beccarianum (Cesati) Diels *Nat. Pfl.* **1**(4): 186 (1899).

Polypodium beccarianum Cesati *Rend. Ac. Sc. Napoli* **16**: 27, 30 (1877).

Dist.: New Guinea.
Specimens: ANDAI: *Beccari s.n.* (FI: K).

Tectaria* cf. *T. brachiata (Zoll. & Moritz) Morton
Ecol.: On forest floor in shade in disturbed lowland rain forest. At 50 m. 1 km from coast.
Specimens: GUNUNG BORAI: Sungai Acemo - *Sands* 6155.
Notes: Fronds to 75 cm tall.

Tectaria cesatiana
Specimens: *Beccari* 12536; *Beccari* 12637.

Tectaria* cf. *T. macrota Holttum
Dist.: Type described from Java. Kepala Burung.
Ecol.: In disturbed lowland forest. On forest floor in shade. 0-20 m alt.
Specimens: ARFAK PLAINS: Wariori flood plain near Masni - *Sands* 6197
Notes: The collection keys out to this species in Flora Malesiana: it is possibly undescribed. Plants to 45 cm tall.

Tectaria cf. ***T. microchlamys*** Holttum
Specimens: *Sands* 6230

***Tectaria* spp.**

1. **Ecol.**: In steep-sided stream in deep shade near river. 70-90 m altitude.
   Specimens: RANSIKI RD.: Near Kali Urkwan. Between Wanmarwai and Warkapi villages - *Sands* 6690.
   Notes: Fronds dimorphic. Plants to 80 cm tall.

2. **Ecol.**: Terrestrial on forest floor in shade in disturbed lowland rain forest.
   Specimens: ARFAK PLAINS: S. Wariori crossing - *Sands* 6255.
   Notes: Plants to 2 m tall with stipes to 70 cm. Fronds with up to 11 pairs of pinnae.

3. **Ecol.**: Terrestrial on forest floor in deep shade in montane forest.
   Specimens: ARFAK MTS.: Mupi Dessa trail - *Sands* 6813.
   Notes: Plants with short trunk to 75 cm. Fronds to 1.5 m.

4. **Ecol.**: Epiphyte in deep shade in montane forest close tp stream.
   Specimens: ARFAK MTS.: Mupi Dessa trail - *Sands* 6855.
   Notes: Erect epiphytic with fronds to 1.5 m with 6-7 pairs of pinnae.

5. **Ecol.**: Terrestrial with clumped rhizome, at 100 - 200 m altitude.
   Specimens: RANSIKI: 2.5 km NW Ransiki - *Johns* 7871.
   Notes: Stipe to 50 cm, base densely covered with scales. Rachis to 70 cm. Pinnae 8-9 + terminal pinna. Individual pinnae to 14x1.3 cm.

6. **Ecol.**: In secondary forest on uplifted coral along road side at 110 m altitude.
   Specimens: G. MEJA: *Johns* 7649.
   Notes: Small stems to 6 cm long. Fronds dimorphic. Sterile fronds: stipe to 30 cm. Fertile fronds: stipe to 40 cm.

7. **Ecol.**: Terrestrial in valley floor forest along track from Wamare to Mokwam at 600-650 m altitude.

Specimens: ARFAK MTS.: Mokwam Trail - *Johns* 7764.
Notes: Caudex short.

8. **Ecol.**: In wet areas and swamps to S of range east of Andjai 2.
Specimens: KEBAR VALLEY: East of Andjai 2 - *Johns* 8283.
Notes: Fronds strongly dimorphic. Sterile fronds: stipe to 33-35 cm, basal pinnae largest - to 23 x 4.3 cm. Fertile fronds: stipe to 47 cm, 15 to 18 pinnae + terminal, to 7 x 0.9 cm. Sori acrostichoid, blackish.

9. **Ecol.**: Terrestrial in ridge top forest on limestone escarpment at 900 m altitude.
Specimens: **KEBAR VALLEY**: On ridge above Apni Valley east of Andjai 2 - *Johns* 8256.
Notes: Rhizome short-creeping, black, covered with scales. Stipe to 60-65 cm long, grooved, with large scales at base. Fronds pinnate.

Specimens: AREA 12: *Johns* 8389. ANDAI: *Johns* 8120. ARFAK PLAINS: *Sands* 6444. KAMPONG MANDOPI - *Sands* 6466.

GLEICHENIACEAE

Holttum, R.E. (1959). Gleicheniaceae. *Flora Malesiana* Series II **1**: 1-36.

DICRANOPTERIS Bernhardi

Dicranopteris linearis (Burm. *f.*) Underw. *Bull. Tor. Bot. Cl.* **34**: 249 (1907). Brass (1941: 283); Holttum (1941: 283); Holttum (1957: 275); Holttum (1959: 33); Nakai (1950: 73); Wagner & Grether (1948: 28).
Polypodium lineare Burm.f. *Fl. Ind.* 255 (1768).
Gleichenia linearis Clarke. Gibbs (1917: 76);
Dist.: Tropics and subtropics.
Ecol.: Climbing trees along forest margins, and also scrambling in forest where more open. 2100 - 2400 m.
Specimens: ARFAK MTS.: *Beccari* s.n. (FI). ANGGI LAKES: *Gibbs* 5575 (BM); *Gibbs* 5991 (BM). *Kanehira & Hatusima* 12916.

Dicranopteris linearis (Burm. *f.*) Underw. var. ***linearis***
Polypodium dichotomum Thunb.
Dist.: Tropical an subtropical Africa to Malesia, Polynesia and Australasia.
Ecol.: Thicket forming but not high climbing, sea-level to 1400 m. In Kebar Valley rare in swampy grassland at 650 m altitude.
Specimens: ARFAK MTS.: Putat - *Beccari s.n.* (FI). KEBAR VALLEY: Anjai 2 - *Johns* 8301.

DIPLOPTERYGIUM Nakai

Diplopterygium longissimum (Blume) Nakaike
Gleichenia glauca (Thunb.) Hook. Gibbs (1917: 76);
Gleichenia longissima Blume

Dist.: Asia to Australia and Polynesia.
Ecol.: Abundant along forest margins. 2100 m.
Specimens: ANGGI LAKES: *Gibbs* 5668 (BM). WITHOUT LOCALITY: *Kanehira & Hatusima* 14156.
Notes: Gibbs records a frond climbing to 7 m in forest.

GLEICHENIA J.E. Smith

Gleichenia vulcanica Blume. Gepp in Gibbs (1917: 76);
Dist.: Malay Islands to New Guinea.
Ecol.: Abundant along forest margins; terrestrial on steep gravel slopes; terrestrial in burnt summit grasslands. Common.
Specimens: ARFAK MTS.: Kostermans 2077. KOEBRE MT. : Gibbs 5611 (BM). ANGGI LAKES: Gibbs 5727 (BM); Gibbs 5996 (BM). WITHOUT LOCALITY: *Kanehira & Hatusima* 13458; *Kanehira & Hatusima* 14057; *van Royen & Sleumer* 7205; *van Royen & Sleumer* 7883.

***Gleichenia* spp.**
1. **Specimens**: IRAY: *Widjaja* 4250.

2. **Specimens**: KP. SUITEBE: *Widjaja* 4378 (BO).

STICHERUS C. Presl

Sticherus bolanicus (Rosenst.) Copeland (1941: 352).
Gleichenia bolanica Rosenst. *Fedde Rep.* **12**: 162 (1913). Alderw. (1917: 86); Holttum (1959: 24). **Type**: Keysser B 14. Mt Bolan, Irian Jaya (B).
Gleichenia subulata Alderw. (1924: 23).
Dist.: New Guinea. Endemic in Irian Jaya.
Ecol.: In open scrub and rocky places especially on limestone. Fronds to 2m long. 2800 - 3950 m.
Specimens: KOEBRE: *Johns* 8060.

Sticherus brassii Nakai *Bull. Nat. Sci. Mus. Tokyo* n. **29**: 15 (1950).
Gleichenia brassii C.Chr. *Brittonia* **2**: 271 (1937). Holttum (1959: 24). **Type**: L.J. Brass 4922. Mt Tafa (Holotype BM: BO).
Dist.: New Guinea: Mts Tafa and Arfak.
Ecol.: Scrambling in open places in forest, 1550-2400 m.
Specimens: ARFAK MTS.: *Kanehira & Hatusima* 14156.

Sticherus hirtus (Blume) Ching
Gleichenia hirta Bl.
Specimens: ARFAK MTS.: *Kostermans* 2093. ANGGI LAKES: *Gibbs* 5577 (BM).

Sticherus truncatus (Willd.) Nakai.
Gleichenia laevigata Hook. Gibbs (1917: 76).
Dist.: Malesia.
Ecol.: Abundant along margins of forest patch.
Specimens: ANGGI LAKES: *Gibbs* 5577 (BM).

GRAMMITIDACEAE

Grammitidaceae spp.

1. **Ecol.**: Epiphyte on trunk over river in montane zone at 850 m altitude.
   Specimens: ARFAK MTS.: Mupi Dessa Trail - *Sands* 6807.
   Notes: Climbing fern to 1 m tall.

2. **Ecol.**: Epiphyte in shade in montane forest at 900 m.
   Specimens: ARFAK MTS.: Mupi Dessa Trail - *Sands* 6828.
   Notes: Erect fronds to 20 cm long.

3. **Ecol.**: Epiphyte in shade in montane forest at 900 m.
   Specimens: ARFAK MTS.: Mupi Dessa Trail - *Sands* 6828.
   Notes: Epiphyte with erect fronds to 20 cm long.

4. **Ecol.**: Epiphyte on trunk in subalpine forest at 1950 m altitude.
   Specimens: KOEBRE MT.: Track from Iray to Mt Koebre - *Johns* 8074.
   Notes: Rhizomes short-creeping. Fleshy to 20 cm long.

5. **Ecol.**: Epiphytic in subalpine forest along path to W summit of Mt Koebre at 1950 m altitude.
   Specimens: ANGGI LAKES: On mountain path near lake in grassland and scrub - *Johns* 8093.
   Notes: Stipe to 3-5 cm. Basal lobes of rachis to 19 cm. Epiphyte in subalpine forest along path to Mountain.

CALYMMODON

Calymmodon cucullatus (Nees et Blume) C. Presl
Polypodium cucullatum Nees et Blume.
Dist.: Ceylon through Malesia to Fiji and New Caledonia.
Ecol.: Epiphyte in forest patches. 2100 m.
Specimens: ARFAK MTS.: *Beccari s.n.* (FI). ANGGI LAKES: *Gibbs* 5734 (BM).

Calymmodon clavifer Hook.) T. Moore
Polypodium clavifer Hook. Gibbs (1917: 75);
Specimens: ARFAK MTS.: *Beccari s.n.* (FI). ANGGI LAKES: *Gibbs* 6119 (BM).

GRAMMITIS Swartz

Parris, B.S. (1983). A taxonomic revision of the genus *Grammitis* Swartz (Grammitidaceae: Filicales) in New Guinea. *Blumea* **29**: 13-222.

Grammitis adspersa (Blume) Blume, *Flora Jarvae* 2: 115 (1830). Copeland 80: 214 (1952); Parris (1983: 44).
Polypodium adspersum Blume *Enum. Pl. Javae* 123 (1828). **Type**: Blume s.n. Java. (L).
Dist.: Southeast Asia, throughout Malesia to Australia and Fiji.
Ecol.: Epiphyte on slender tree trunks in lower montane and mid montane forest.
Specimens: ANGGI LAKES: Anggi Gita - Kostermans 2221 (BO).
Notes: Parris (1983: 43) lists several synonyms for this species (None from Irian Jaya).

Grammitis ahenobarba Parris, *Blumea* **29**(1): 142 (1983). **Type**: L.S. Gibbs 5551. -/12/1913. Anggi Lakes, Vogelkop Peninsula. (BM).

Dist.: Local endemic near Anggi Lakes in the Kepala Burung.

Ecol.: Epiphyte in montane forest at 2130 m.

Specimens: ANGGI LAKES: Anggi Gita - Gibbs 5551 (BM).

Grammitis dolichosora (Copel.) Copel. *Philip. J. Sc.* **80**: 188 (1952). Parris (1983: 191).

Polypodium dolichosorum Copel. *Philip. J. Sc.* **1**: Suppl. 159 (1906). **Lectotype**: Copeland 1524. Davao, Mindanao, Philippines. (MICH: K, UC).

Dist.: Borneo, Philippines and New Guinea.

Ecol.: A low epiphyte on trunks, and roots, of trees, sometimes in moss cushions. Occasionally terrestrial or rupestral.

Specimens: ANGGI LAKES: Anggi Gita - Gibbs 5972 (BM); Kostermans 2210 (BO); Kostermans 2211 (BO).

Grammitis fasciata Blume

Polypodium fasciatum (Blume) C.Presl. Gibbs (1917: 74);

Dist.: Malesia.

Ecol.: Epiphyte in forest. 2100 m.

Specimens: ANGGI LAKES: Gibbs 5971 (BM);

Grammitis knutsfordiana (Baker) Copel. *Univ. Calif. Publ. Bot.* **18**: 224 (1942). Copel. (1952: 201).

Polypodium knutsfordianum Baker *J. Bot.* (London) **28**: 107 (1890). **Type**: Macgregor 29 (K).

Polypodium oleandroides Baker *Ann. Bot.* **8**: 128 (1894). **Type**: Macgregor s.n. Mt Suckling. 1892 (K).

Polypodium diplosoroides Rosenst. *Nova Guinea* **8**: 724 (1912). Copel. (1952: 187). **Lectotype**: von Romer 1167 (BO).

Polypodium fasciatum sensu. Ridley *Trans. Linn. Soc. London* (Bot.) **2**(9): 259 (1916), quoad Boden Kloss spec.

Polypodium diplosorum var. sensu Brause (1920: 180), quoad Ledermann 11781.

Polypodium hookeri sensu Alderw. (1924: 44), quoad Lam 1941.

Dist.: Sumatra, Borneo and Philippines to New Guinea, Solomon Islands and Fiji.

Ecol.: Usually a more or less erect epiphyte on tree trunks or logs. Montane species, rarely subalpine. 760 to 3500 m altitude.

Specimens: MT NETTOTI: van Royen & Sleumer 8181. (L).

Grammitis padangensis (Baker) Copel. *Philip. J. Sc.* 80: 212 (1952). Parris (1983: 85).

Polypodium padangense Baker *J. Bot.* (London) **18**: 213 (1880). **Type**: Beccari s.n. Sumatra, Mt. Singalan. 1700 m. (K).

Polypodium stanleyanum Baker *J. Bot.* (London) **28**: 107 (1890). Grammitis stanleyana Copel. (1952: 168). **Type**: MacGregor 25. (K: BM).

Polypodium trichocarpum Alderw. *Nova Guinea* **14**: 41 (1924). **Lectotype**: Lam 1683 (L: BM).

Polypodium trichocarpum Alderw. var. ***inerme*** Alderw. *Nova Guinea* **14**: 42 (1924). **Lectotype**: Lam 1854 (L: BM).

Dist.: From Sumatra, Borneo and Philippines to New Guinea and Bismark Archipelago.

Ecol.: Epiphyte on tree trunks and branche, often in subalpine forest and montane forest. Occasionally epiphytic on *Cyathea*. Sometimes terrestrial or rupestral in high altitude forest, occasionally rupestral in subalpine grassland. Altitudinal range from 1830 to 3880 m.
Specimens: MT NETTOTI: van Royen & Sleumer 8004. (L).

Grammitis reinwardtii Blume, *Enum. Pl. Javae Addend.* (1828). Copel. (1952: 236); Parris (1983: 107).

Polypodium reinwardtii C. Presl *Tentamen* 180 (1836). **Lectotype**: Reinwardt *s.n.* G. Klabat, Celebes. (L).

Polypodium sucklingianum Baker *Ann. Bot.* **8**: 128 (1894). Copel. (1952: 187). *Grammitis sucklingiana* Parris (1980: 118). **Type**: Macgregor s.n. Mt Suckling, Papua New Guinea. (K).

Polypodium trichopodum var. ***serratolobatum*** Brause *Bot. Jahrb.* **50**: 179 (1920). **Lectotype**: Ledermann 11576 (B).

Grammitis mariae Copel. *Univ. Calif. Publ. Bot.* **18**: 223 (1942). **Type**: Clemens 10251a. (MICH).

Grammitis matapensis Copel. *Univ. Calif. Publ. Bot.* **18**: 223 (1942). **Type**: Clemens 11176. (MICH: BM, UC).

Dist.: Ceylon and mainland SE Asia, throughout Malesia to Solomon Islands and Australia.
Ecol.: Usually low epiphyte, sometimes on *Cyathea*. Rarely on rocks in stream beds. Lower montane and mid montane forest. From c. 400 to 3660 m altitude.
Specimens: NETTOTI RANGE: van Royen & Sleumer 7969. (L). **WONDIWOI MTS.**: Mayr 316. (BO).

Grammitis subfasciatua (Rosenst.) Copel., *Philip. J. Sc.* 80: 170 (1952). Parris (1983: 83).

Polypodium subfasciatum Rosenst. *Feddes Repert.* **5**: 41 (1908). **Type**: Werner 25. (UC: S).

Polypodium integrum Brause *Bot. Jahrb.* **49**: 37 (1912). *Grammitis integra* Copel. (1952: 171). **Lectotype**: Schultz Jena (26) 21 (B).

Polypodium fuciforme Rosenst. *Nova Guinea* **8**: 726 (1912). Copeland (1952: 203). **Type**: von Romer 846 (BO).

Polypodium alcicorne Ridley *Trans. Linn. Soc. London* (Bot.) **2**(9): 261 (1916). non Baker (1888). Copel. (1952: 204). **Lectotype**: Boden Kloss s.n. Mt Carstensz. Camps III to VIII. 750 to 1500 m. (BM: K).

Polypodium dichotomum Brause *Bot. Jahrb.* **56**: 190 (1920) non Houtt. (1783) nec Thunb. (1784). Copeland (1952: 204). **Lectotype**: Ledermann 9246 (B: BM, S).

Dist.: Moluccas, throughout New Guinea to Morobe and Central Province.

Ecol.: Erect (rarely pendulous) epiphyte in tree crowns, or in moss or rotten logs. Occasionally rupestral. Usually lower montane to midmontane forests (*Nothofagus/Podocarpus*), rarely in lowlands. From 50 - 2590 m.

Specimens: ANGGI LAKES: Anggi Gita - Gibbs 5971 (BM).

Grammitis sumatrana (Baker) Copel. Parris (1983: 49-53) does not record this species from the Kepala Burung.

PROSAPTIA C.Presl

Prosaptia contigua (G.Forst) C. Presl. Gepp in Gibbs (1917: 73);

Davallia contigua G. Forst.

Dist.: Malesia to Polynesia. In Irian Jaya - Arfak Mts., Noord River, Mt Jaya.

Ecol.: Epiphyte in isolated forest patch. 2100 m.

Specimens: ANGGI LAKES: *Gibbs* 5895 (BM).

Prosaptia schlechteri C.Chr. Gepp in Gibbs (1917: 72);

Davallia schlechteri C.Chr.

Dist.: New Guinea. Arfak Mts. to Mt Jaya.

Ecol.: Epiphyte in montane forest on slopes. 2400 - 2700 m.

Specimens: KOEBRE MT.: *Gibbs* 5634 (BM); *Gibbs* 5640 (BM); *Gibbs* 5625 (BM).

XIPHOPTERIS Kaulf.

Xiphopteris subpinnatifida Blume.

Ecol.: Epiphyte in forest.

Specimens: ARFAK MTS.: *Kostermans* 2221 (BO).

***Xiphopteris* sp.**

1. **Ecol.**: Epiphyte in shade in montane forest at 850 m altitude.
   Specimens: ARFAK MTS.: Mupi Dessa Trail - *Sands* 6857.
   Notes: Epiphyte with fronds to 45 cm long.

HYMENOPHYLLACEAE

Trichomanaceae Kunkel [Trichomanaeaceae orthog. err.]

Copeland, E.B. (1933). *Trichomanes*. *Philip. J. Sci.* **51**: 119 - 280.

Copeland, E.B. (1937). *Hymenophyllum*. *Philip. J. Sci.* **64**: 1 - 118.

Copeland, E.B. (1938). Genera Hymenophyllacearum. *Philip. J. Sci.* **67**: 1 - 110.

Croxall, J.P. (1975). The Hymenophyllaceae of Queensland. *Aust. J. Bot.* **23**: 509 - 547.

Iwatsuki, K. (1975). Studies in the systematics of filmy ferns. I. A note on the identity of *Microtrichomanes*. *Brit. Fern Ga.* **11**: 115 - 124.

Iwatsuki, K. (1975). Studies in the systematics of filmy ferns. II. A note on *Meringium* and the taxa allied to this. *Gard. Bull. Sing.* **30**: 63 - 74.

Morton, C.V. (1968). The genera, subgenera and sections of the Hymenophyllaceae. *Contr. U.S. Nat. Herb.* **38**: 153-214.

A widsespread,distinctive, predominantly tropical and southern temperate family with up to 40 genera (or 2 according to some authorities), containing over 400 species. In Papuasia there are some c. 100 species in 20 genera.

HYMENOPHYLLACEAE spp.

1. **Ecol.**: Epiphyte on dead trunk in partial shade. In lower montane forest at 775 m.
Specimens: ARFAK PLAINS: On steep ridges to south to G. Itsiwei - *Sands* 6390.
Notes: Mat forming.

2. **Ecol.**: In disturbed hill forest in partial shade near stream. Epiphyte on fallen tree trunk at 395 m.
Specimens: ARFAK MTS.: Mupi Dessa Trail (Track to G. Humibou) - *Sands* 6873.
Notes: Erect epiphyte with fornds to 40 cm long.

3. **Ecol.**: Epiphyte in shade. In lower montane forest at 800 m.
Specimens: ARFAK PLAINS: On steep ridges to south to G. Itsiwei - *Sands* 6389.
Notes: Fronds erect to 15 cm, stipes dark blue-green.

4. **Ecol.**: Terrestrial on ridge forest on wetter slopes in lower montane forest at 800-850 m.
Specimens: ARFAK MTS.: Mokwam Trail - Johns 7687.
Notes: On wet slopes. Caudex to 1-2 cms.

5. **Ecol.**: Terrestrial in ridge forest on wetter slopes in lower montane forest at 800-850 m.
Specimens: ARFAK MTS.: Mokwam Trail - Johns 7692.
Notes: Caudex to 2-3 cms. Fronds pinnate.

6. **Ecol.**: Growing densely on trunk of rotting log on ground. In lower montane forest at 800-850 m.
Specimens: ARFAK MTS.: Mokwam Trail - Johns 7700.
Notes: Plants to 10 cm. Rachis broadly winged to 1 cm wide, lowest 1 cm unwinged. Lamina is very narrow (often less than wing). Rachis wing with scales.

7. **Ecol.**: Small terrestrial fern very common along valley sides in wet lower montane forest at 800-850 m.
Specimens: ARFAK MTS.: Mokwam Trail - Johns 7744.

Notes: Plants with trunk to 10 cm. Stipe to 6 cm densely covered with small scales. Rachis to 21cm.

8. **Ecol.**: Epiphyte in ridge forest on wetter slopes in lower montane forest dominated by *Araucaria* mixed with *Lithocarpus* and *Litsea* at 650 m altitude.
Specimens: ARFAK MTS.: Mokwam Trail - Johns 7789.
Notes: Stipe winged to base, margins ttothed. Lamina reduced to a series of teeth, joined at base by a wing. Sori terminal.

9. **Ecol.**: Epiphyte on trees in ridge top forest along limestone escarpment at 900 m altitude.
Specimens: KEBAR VALLEY: E of Andjai 2 - *Johns* 8270.
Notes: Rhizome long-creeping, wiry, black. Stipes to 1 cm long, black. Fronds entire to 1.5 x 1 cm.

10. **Ecol.**: Epiphyte on low tree trunks in ridge top forest along limestone escarpment at 900 m altitude.
Specimens: KEBAR VALLEY: E of Andjai 2 - *Johns* 8274.
Notes: Rhizome stout, long-creeping, stipes to 2 cm, winged in upper half. Margins entire.

11. **Ecol.**: Lithophytic, long-creeping rhizome spreading over surface of rocks in swamp forest in dark shade at 600 m altitude.
Specimens: KEBAR VALLEY: E of Andjai 2 - *Johns* 8251.
Notes: Rhizome long-creeping, black. Stipes to 1 cm. Margins entire, to 3.2 x 2.3 cm.

12. **Ecol.**: Lithophytic, long-creeping rhizome spreading over surface of rocks in swamp forest at 600 m altitude.
Specimens: KEBAR VALLEY: E of Andjai 2 - *Johns* 8253.
Notes: Rhizome long-creeping, black. Stipes to 1 cm, winged. Margins entire, to 2 x 1.5 cm.

13. **Ecol.**: Epiphyte on trunk in subalpine forest at 1950 m altitude.
Specimens: MT KOEBRE: Along path from Iray to summitt - *Johns* 8078.
Notes: Mixed collection of terrestrial species.

14. **Ecol.**: Epiphyte on trunk in subalpine forest at 1950 m altitude.
Specimens: MT KOEBRE: Along path from Iray to summitt - *Johns* 8082.
Notes: Fronds small to 2 cm long. Margins serrate.

15. **Ecol.**: Epiphyte on trunk in subalpine forest at 1950 m altitude.
Specimens: MT KOEBRE: Along path from Iray to summitt - *Johns* 8076.
Notes: Rhizome very fine, long-creeping. Stipe to 4.5 cm long. Pinnate, fronds silvery-grey.

16. **Ecol.**: Epiphyte on trunk in subalpine forest at 1950 m altitude.
Specimens: MT KOEBRE: Along path from Iray to summitt - *Johns* 8083.
Notes: Plants small, branching in a single directio. Margins entire.

17. **Ecol.**: Epiphyte on trunk in subalpine forest at 1950 m altitude.
Specimens: MT KOEBRE: Along path from Iray to summitt - *Johns* 8084.

18. **Ecol.**: Terrestrial on banks of stream and wetter patches on lower slopes at 300 m altitude.
Specimens: ANDAI FORESTS: In mature forest on uplifted coral - *Johns* 8112.
Notes: Rhizome forming a small trunk to 4-5 cm all. Plants to 10 cm. Fronds very fine.

19. **Ecol.**: Terrestrial on banks of stream and wetter patches on lower slopes at 300 m altitude.
Specimens: ANDAI FORESTS: In mature forest on uplifted coral - *Johns* 8113.
Notes: Rhizome forming a small trunk to 5-6 cm all. Plants to 15 cm. Fronds translucent, very pale green.

20. **Ecol.**: Epiphyte on tree trunks in valley forest at 300 m altitude.
Specimens: ANDAI FORESTS: In mature forest on uplifted coral - *Johns* 8115.
Notes: Stipes to 1-1.5 cm long, not winged. Rachis to 1.5 cm long. Margins of fronds entire.

21. **Ecol.**: Epiphyte on tree trunks in valley forest at 300 m altitude.
Specimens: ANDAI FORESTS: In mature forest on uplifted coral - *Johns* 8118.
Notes: Rhizome creeping. Stipes to 1cm long, slightly winged in upper parts. Fronds to 2-4 cm long. of fronds entire.

22. **Ecol.**: Epiphyte with climbing rhizome on tree trunks in disturbed lowland rain forest at 50 m altitude.
Specimens: ARFAK PLAINS: Wariori Flood Plains - *Johns* 8126.
Notes: Rachis with many small fronds scattered along it. Fronds to 4-5 cm long. Stipes to 2.6 cm.

23. **Ecol.**: Epiphyte with climbing rhizome on tree trunks in disturbed lowland rain forest at 50 m altitude.
Specimens: ARFAK PLAINS: Wariori Flood Plains - *Johns* 8145.
Notes: Fronds pendant. Stipes to 20 cm long. Rachis to 65 cm.

24. **Ecol.**: Growing on rock surfaces along streams in river valley in lower montane forest to W of camp at 80- 150 m altitude.
Specimens: ARFAK PLAINS: Wariori River Valley - *Johns* 8204.
Notes: Plants to 2-5 cm small. Stipes very short with winged rachis.

25. **Ecol.**: Terrestrial in ridge forest on wetter slopes in lower montane forest dominated by *Araucaria*, *Lithocarpus* and *Litsea*. 650 m altitude.
Specimens: ARFAK MTS.: Mokwam Trail - *Johns* 7787.
Notes: Stipe to 8.5 cm long. Rachsi to 15 x 2.5-3 cm, winged, brown.

26. **Ecol.**: Terrestrial in ridge forest on wetter slopes in lower montane forest dominated by *Araucaria*, *Lithocarpus* and *Litsea*. 650 m altitude.
Specimens: ARFAK MTS.: Mokwam Trail - *Johns* 7803.
Notes: Trunk to 8-10 cm tall. Stipe 5-6 cm. Fronds pinnate. rachis to 18-19 cm long with c. 30 pinnae.

27. **Ecol.**: Terrestrial in ridge forest on wetter slopes in lower montane forest dominated by *Araucaria*, *Lithocarpus* and *Litsea*. 650 m altitude.
Specimens: ARFAK MTS.: Mokwam Trail - *Johns* 7804.

Notes: Trunk to 4-5 cm tall. Fronds bipinnatifid. Stipes not winged.

28. **Ecol.**: Epiphyte on tree in Hot Springs area at 600 m altitude.
Specimens: KEBAR VALLEY: S of Andjai 2 - *Johns* 8368.
Notes: Rhizome long-creeping, narrow, black. Stipe to 1.5 cm long, with scattered hairs. Rachsi to 3.5 cm, broadly winged. Margins entire.

29. **Ecol.**: Epiphyte on buttress of *Ficus* in Hot Springs area at 600 m altitude.
Specimens: KEBAR VALLEY: S of Andjai 2 - *Johns* 8366.
Notes: Rhizome long-creeping, thick, covered with brown scales. Stipe to 1-1.3 cm, scaly at base. Fronds to 2 cm long.

CEPHALOMANES C. Prel

Cephalomanes* cf. *javanicum (Bl.) Copel
Ecol.: On steep mud bank along gully side on steep ridge in hill forest.
Specimens: ARFAK MTS.: Mokwam Trail - *Sands* 6142.
Notes: Fronds tufted at apex of short trunk. Fronds very dark green.

Cepahalomanes javanicum v.d.Bosch. **Type**: Tindale (1963: 47).
Notes: *C. atrovirens* stipes are 1 - 4.5 cm in length but are 8 - 12 cm long in *C. javanicum*.

***Cephalomanes* sp.**
1. **Ecol.**: On mud banks near river in partial shade. Monatane forest at 900 m.
Specimens: ARFAK MTS.: Mupi Dessa Trail - *Sands* 6830.
Notes: Erect to 30 cm tall. Stipes dark-olive green. Fronds dark green-blue.

HYMENOPHYLLUM L.

Hymenophyllum australe Willd. L. Sp. Pl. ed. 4. 5 (1810) 527. Brause (1920:40); Gepp. in Gibbs (1917: 68); Tindale (1963: 25); Alderw. (1909: 69). **Type**:
Dist.: SE Asia: India, Malesia [N Borneo, Lesser Sunda Is.] to Australia (Qld. NSW Vic. Tas.) New Zealand and Antarctic Islands. In Papuasia: W NE New Guinea. From India, throughout Malesia to Australasia. In Irian Jaya: Kepala Burung.
Ecol.: Epiphyte in forest patches near lake. 2100 m.
Specimens: ANGGI LAKES: Gibbs 5893 (BM). ARFAK MTS.: Beccari sn.
Notes: Tindale (1963) states that this species occurs only in Australia and New Zealand.

Hymenophyllum cernuum Gepp. in Gibbs, *Phytogeog. and Flora Arfak Mts.* 68 (1917). **Type**: L.S. Gibbs 5964. 12/1914. Anggi Lakes, near Female Lake. Epiphyte in forest patch. 2100m. (Holotype: BM)
Dist.: Endemic.
Ecol.: Epiphye in forest patches near lake. 2100 m.
Specimens: ANGGI LAKES: Gibbs 5964 (Holotype: BM)
Notes: Gepp. notes its position in *Leptocionium*. General hairiness of frond and the narrowly winged stipe and rachis are characteristic. Otherwise similar to *H. zollingerianum* (v.d.Bosch, *Hymen. Javan.* t.50).

Hymenophyllum cincinnatum Gepp. in Gibbs, *Phytogeog. and Flora Arfak Mts.* 68 (1917). **Type**: L.S. Gibbs 5989. 12/1914. Anggi Lakes. Epiphyte in moss-forest. 2550m. (Holotype: BM)

Dist.: Endemic.

Ecol.: Epiphyte in moss-forest. 2550 m.

Specimens: ANGGI LAKES: Gibbs 5989 (BM).

Hymenophyllum javanicum Spreng var. ***atrovirens*** Colenso. Beccari (1886: 32); Cesati (1877: 24).

Dist.: West New Guinea.

Specimens: ARFAK MTS.: Beccari s.n.

Notes: Cesati notes that this species is very close to *Hymenophyllum atrovirens*.

Hymenophyllum kurzii Prantl. Brause (1920: 46); Gepp in Gibbs (1917: 68).

Dist.: Java, NW New Guinea.

Ecol.: Epiphyte in forest patches near lake.

Specimens: ANGGI LAKES: Gibbs 5679 (BM).

Hymenophyllum paniculiflorum Presl. Gepp in Gibbs (1917: 68);

Dist.: Japan to Malesia. In Irian Jaya in NW.

Ecol.: Epiphyte in forest patches near lake.

Specimens: ANGGI LAKES: Gibbs 6141 (BM).

Hymenophyllum pediculariaefolium Cesati *Rendic. Acad. Sc. Fis. Mat. Napoli.* **16**: 24, 28 (1877). Beccari (1886: 32); Brause (1920: 40); Alderw. (1909: 74). **Type**: *Beccari s.n.* ex Mt Arfak ad Hatum, Luglio 1875.

Dist.: W New Guinea.

Specimens: MT ARFAK: Hatum - Beccari *s.n.*

Notes: As '*pediculariaefolium*'.

Hymenophyllum puellum Cesati *Rendic. Accad. Sc. Fis. Mat. Napoli* **16**: 24, 28 (1877). Type: Beccari s.n. ex Mt Arfak at Hatam, Luglio 1875.

Hymenophyllum capillaceum Roxb.

Trichomanes apiifolium Beccari (1886) 32.

Specimens: MT ARFAK: Hatam - Beccari s.n.

Hymenophyllum salakense Racib.

Dist.: Java and Irian Jaya.

Ecol.: Epiphyte in forest patches. 2100 m.

Specimens: ANGGI LAKES: Gibbs 6138 (BM);

Hymenophyllum **spp.**

1. **Ecol.**: Epiphyte on tree trunk in *Castanopsis* forest at 1550 m.
Specimens: KEBAR VALLEY: Tohkiri Mts. Aifat River Valley - *van Royen & Sleumer* 7261 (L).

2. **Ecol.**: Along trail on mossy floor in *Nothofagus* forest at 2100 m. Very common terrestrial.
Specimens: ANGGI LAKES: Mt Sarumot, Iray - *van Royen & Sleumer* BW 14328 (L).
Notes: Stipe densely hairy.

3. **Ecol.**: On mossy floor in *Nothofagus* forest at 2100 m.

Specimens: KEBAR VALLEY: Mt Nettoti along path from Andjai to Wekari River - *van Royen & Sleumer* 7920 (L).
Notes:

4. **Ecol.**: In forest on mossy crest at 2100 m
Specimens: KEBAR VALLEY: Summit of Mt Nettoti - *van Royen & Sleumer* 7949 A (L).
Notes: Rare.

5. **Ecol.**: In low forest at 2100 m.
Specimens: ANGGI LAKES: *Kanehira & Hatusima* 13454 (BO).

6. **Ecol.**: Epiphyte in forest at 1800 m.
Specimens: ARFAK MTS.: *A.J.G.H. Kostermans* 2244. (BO).

7. **Ecol.**: Epiphyte in forest at 1800 m.
Specimens: ARFAK MTS.: *A.J.G.H. Kostermans* 2377. (BO).
Notes: ***Pleuromanes*** sp.

8. **Ecol.**: In low forest near lake at 2100 m.
Specimens: ANGGI LAKES: *Kanehira & Hatusima* 14071 (BO).

9. **Ecol.**: In low forest above lake at 300 m.
Specimens: ANGGI LAKES: *Kanehira & Hatusima* 12440 (BO).

10. **Ecol.**: Epiphyte in forest at 1900 m.
Specimens: ARFAK MTS.: *A.J.G.H. Kostermans* 2137. (BO).

11. **Ecol.**: On path to summit at 2000 m
Specimens: KEBAR VALLEY Mt Nettoti - *van Royen & Sleumer* 7992 (BO).

12. **Ecol.**: In forest on mossy crest at 1950 m.
Specimens: KEBAR VALLEY: Summit of Mt Nettoti - *van Royen & Sleumer* 7993 (BO).

13. **Ecol.**: On path to summit at 1950 m
Specimens: KEBAR VALLEY Mt Nettoti - *van Royen & Sleumer* 8217 (BO).

TRICHOMANES L.

Trichomanes aphlebioides Christ. Gibbs (1917: 67)
Dist.: NW New Guinea.
Ecol.: Epiphyte in forest near lakes. 2100 m.
Specimens: ANGGI LAKES: Gibbs 5947 (BM); Gibbs 6137 (BM). BEGOWRI RIVER: Gjellerup s.n.

Trichomanes apiifolium Presl Annon. (1899: 118); Baker (1890: 105); Beccari (1886: 34); Cesati (1877: 25); Schlechter in Brause (1912: 3); K.Sch. & Laut. (1905: 36).
Dist.: Malaysia to Polynesia and Norfolk Is. In Papuasia: W NE New Guinea.
Ecol.: Epiphyte in hill forest at 600 to 1000 m altitude.
Notes: ***Trichomanes maximum*** (Beccari 1886). Section Eutrichomanes (K.Sch. & Laut. 1905: 36).

Trichomanes cognatum Cesati *Rendic. Acad. Sc. Fis. Mat. Napoli* **16**: 24, 28 (1877). Beccari (1886: 33). **Type**: Beccari s.n. Andai, Arfak Mts. (FI).
Dist.: Arfak Mts.
Specimens: ARFAK MTS.: Andai - *Beccari* s.n. (FI).

Trichomanes digitatum Sw. Gibbs (1917: 67);
Dist.: Malesia to Mascarene islands and E Australia.
Ecol.: Epiphyte in moss forest. 1800 m.
Specimens: ANGGI LAKES: Gibbs 6000 (BM);

Trichomanes meifolium Bory. Gibbs (1917: 67);
Dist.: Malay Peninsula to Polynesia and Reunion.
Ecol.: Epiphyte in wet forest. 2100 m.
Specimens: ANGGI LAKES: Gibbs 5678 (BM). ARFAK MTS.: Beccari sn. (FI). WITHOUT LOCALITY: BW 7982.

Trichomanes pallidum Bl. Gibbs (1917: 67)
Dist.:From tropical Asia to Polynesia. In Irian Jaya from Kepala Burung, Mt Hellwig and Mt Jaya.
Ecol.: Epiphyte in forests. 2400 - 2700 m.
Specimens: MT. KOEBRE: Gibbs 5728 (BM); ARFAK MTS.: Hatam - Beccari s.n. (FI);

Trichomanes palmatifidum K. Mull. Gibbs (1917: 67);
Dist.: Java, Irian Jaya (Vogelkop).
Ecol.: Epiphyte in moss forest. 2400 m.
Specimens: ANGGI LAKES: Gibbs 5520 (BM);

LINDSAEACEAE

Kramer, K.U. (1971). Lindsaea-Group. *Flora Malesiana* Ser. II. Vol. **1**(3): 177- 254.

LINDSAEA Dryander

Lindsaea cultrata (Willd.) Swartz *Syn. Fil.* 119 (1806).
Dist.: Throughout Malesia to NE Australia. Apparently uncommon in New Guinea.
Ecol.: In gully In rain forest at 125 - 160 m altitude.
Specimens: ANDAI: *Beccari s.n.* (FI). RAMOI: *Beccari s.n.* (FI).

Lindsaea kingii Copel.
Dist.: Moluccas and New Guinea to Solomon Islands.
Ecol.: Terrestrial in ridge top forest along limestone escarpment at 900 m altitude.
Specimens: KEBAR VALLEY: E of Anjai 2, Apni Valley - *Johns* 8267.
Notes: Stipes to 30-35 cm, greyish-brown. Fronds bipinnate.

Lindsaea lobata Poiret in Lamk. *Encycl. Suppl.* **3**: 448 (1813). Kalkman (1971: 219) **Type**: *Commerson s.n.* Java (Holotype P).
Dist.: Throughout Malesia. Distinct varieties in Hainan and Indo-China.
Ecol.: Terrestrial or on tree trunks in mature forest. (50-)600-1800 m. Epiphyte on dead trunk in partial shade. In lower montane forest at 775 m.
Specimens: ANDAI: *Beccari s.n.* (FI).

Lindsaea lucida Bl. *Enum. Pl. Jav.* 216 (1828). Holttum (1954: 328); Kramer (1968: 567); Kramer (1971: 233). **Type**: *Blume s.n.* Java (L).

Specimens: PRAFI: *Widjaja* 3164 *n.v.* (BO).

Notes: Kramer (1968; 1971: 234) recognises two subspecies. Subspecies *lucida* is rare in New Guinea.

Lindsaea lucida Bl. var. ***brevipes*** (Copeland) Kramer *Blumea* **15**: 567 (1968).

Lindsaea brevipes Copeland *Philip. J. Sci. Bot.* **6**: 83 (1911).

Dist.: From Moluccas to New Guinea, including Admiralty Islands and Solomon Islands.

Ecol.: Terrestrial on rocks, rarely an epiphyte. Scattered in rainforest from sea-level to 400 - 1600 m. Common on forest floor of ridge top lower montane forest in shade. 675 m altitude.

Specimens: ARFAK PLAINS: Ridge to G. Itsiwei - *Sands* 6426. WITHOUT LOCALITY: *Beccari* 12386; *Beccari* 12387. TOR RIVER: *Gjellerup* 752 (L).

Notes: Plants to 35(-40) cm. Stipe and rachis dark maroon.

Lindsaea microstegia Copeland *Philip. J. Sci. Bot.* **6**: 83 (1911). Copeland (1949: 20); Kalkman (1971: 246). **Type**: *King* 242, Papua (Holotype MICH).

Dist.: Throughout New Guinea to D'Entrecasteaux Is.).

Ecol.: Epiphytic, rarely terrestrial. In moist forest from 80-2100 m alt.

Specimens: BW 10327. *van Royen & Sleumer* 6890.

Notes: Very closely related to *L. versteegii* (Christ) Alderw. and *L. rosenstockii* Brause.

Lindsaea obtusa J.Sm. in Hook. *Sp. Fil.* **1**: 224 (1846). Kalkman (1971: 218). **Type**: Cuming 394. Malacca. (Holotype K: B, E, GH, P, W).

Dist.: Taiwan, throughout Malesia to Solomon Is. and Queensland.

Ecol.: In forest on banks, rocks and tree trunks. Sea-level to 1500(-2000) m. Common on forest floor of lower montane forest in shade. 600-850 m altitude.

Specimens: ARFAK PLAINS: Ridge to G. Itsiwei - *Sands* 6377. Wariori flood plain - Sands 6210. KEBAR VALLEY: Anjai 2 - *Johns* 8300. WITHOUT LOCALITY: BW 6359.

Notes: Plants to 20-30 cm. Stipe and rachis dark chocolate-brown.

Lindsaea odorata Roxb. *Calc. J. Nat. Hist.* **4**: 511 (1844). Kramer (1968: 567); Kramer (1971: 228). **Type**: *Icones Roxburghianae* pl. 2578 (K).

Lindsaea crassipes Rosenst. *Fedde Repert* **5**: 36 (1908). **Type**: Werner 74. Mojo R. New Guinea (Holotype B: L, US).

Dist.: Widespread from Zimbabwe to China and Japan. Throughout Malesia to Solomon Is.

Ecol.: On banks from sea-level to 2000 m. Locally common throughout most of its distribution. Rarer in New Guinea and Solomon Islands.

Specimens: KEBAR VALLEY: *van Royen & Sleumer*5062.

Notes: *L. crassipes* is dwarfed, with rigid pinnules and very dark axes.

Lindsaea repens (Bory) Thwaites

Lindsaea hymenophylloides Bl.

Dist.: Malesia to Polynesia.

Ecol.: Epiphyte in moss forest. 2400 m.

Specimens: ANGGI LAKES: *Gibbs* 6120 (BM).

Lindsaea repens (Bory) Thwaites var. sessilis (Copel.) Kramer.

Ecol.: Climbing epiphyte to 4 m in hill forest.

Specimens: ARFAK MTS.: Muppi Dessa (NE of Siembokwon Village) - *Sands* 6881. KEBAR VALLEY: Hot Springs - *Sands* 6622.

Notes: Rhizomes coffee brown and covered with scales.

Lindsaea rigida J.Sm.
Dist.: Malesia.
Ecol.: Epiphyte and terrestrial in forest on slopes. 2250 - 2550 m.
Specimens: KOEBRE MT.: *Gibbs* 5626 (BM). KEBAR VALLEY: Hot Springs - Johns 8374. WITHOUT LOCALITY *van Royen & Sleumer* 7942. *van Royen & Sleumer* 8180. *van Royen & Sleumer* 8182.

Lindsaea tenuifolia Blume *Enum. Pl. Jav.* 219 (1828). **Type**: Blume s.n. Java. (L).
Dist.: Sulawesi, Moiluccas, Sumatra, Java to New Guinea.
Ecol.: Terrestrial along small valley in mature rain forest on uplifted coral. To 1000 m altitude. On weathered limestone on forest floor. Partially disturbed lowland rain forest.
Specimens: ANDAI: 2.5 km W of Andai - *Johns* 8105. PRAFI: *Widjaja* 3163. ARFAK PLAINS: Andai Forest Reserve - *Sands* 6891. WITHOUT LOCALITY: *van Royen & Sleumer* 6915.
Notes: Plants to 30 cm. Stipe and rachis yellowish green.

***Lindsaea* spp.**

1. **Ecol.**: Terrestrial in ridge forest in mid-montane forest at 800-850 m altitude.
   Specimens: **ARFAK MTS.**: Mokwam Trail - *Johns* 7691.

2. **Ecol.**: Growing in dense patches on fallen logs in ridge top frorest dominated by *Araucaria, Lithocarpus* and *Litsea* at 650 m altitude.
   Specimens: **ARFAK MTS.**: Mokwam Trail - *Johns* 7785.

3. **Ecol.**: Growing in dense patches on fallen logs in ridge top frorest dominated by *Araucaria, Lithocarpus* and *Litsea* at 650 m altitude.
   Specimens: **ARFAK MTS.**: Mokwam Trail - *Johns* 7788.

Specimens: ARFAK PLAINS: Mountains to South - *Sands* 6426. ARFAK MTS.: Mokwan Trail - *Johns* 7785; *Johns* 7788. KEBAR VALLEY: Anjai 2 - *Johns* 8300. WITHOUT LOCALITY: *Croxall* 24; *Sands* 6210.

SPHENOMERIS Maxon

Sphenomeris chinensis (L.) Maxon *J. Wash. Ac. Sc.* **3**: 144 (1913). Kalkman (1968: 572); Kalkman (1971: 182). **Type**: *Osbeck s.n.* China (S-PA).

var. ***chineensis***

Dist.: Throughout tropics and subtropics of Old World, not in continental Africa.
Ecol.: Terrestrial and on rocks in thickets and in open forest. From (100-)800-1500(-2400) m.
Specimens: ARFAK MTS.: Kostermans 2294 (L). **ANGGI LAKES**: Anggi Gita - Bivouac Noordpool - *Sleumer & Vink* BW 14027 (L).

Sphenomeris retusa (Cav.) Maxon *J. Wash. Ac. Sc.* **3**: 144 (1913). Kalkman (1971: 180).

Davallia retusa Cav. *Descr.* 278 (1802). **Type**: *Nee s.n.* Mt. Isarrog, Luzon. (Holotype MA: Photo U).

Dist.: Malesia to Solomon Is. and Vanuatu.
Ecol.: On banks in open places, often amongst rocks by river, less often in forest. From sea-level to c. 2000 m altitude.
Specimens: *Beccari* 12359 (FI).

***Sphenomeris* sp.**

1. **Ecol.**: Terrestrial near swamp on dry hillside above lake.
Specimens: ANGGI LAKES: *Johns* 8099.
Notes: Stipe to 50 cm long. Rachis of young plants reddish. Sori at apex of small pinnae.

TAPEINIDIUM (C. Presl) C. Chr.

Tapeinidium longipinnulum (Ces.) C.Chr. *Ind. Fil. Suppl.* **3**: 176 (1934). Kramer (1968: 555); Kramer (1971: 197).

Davallia longipinnula Ces. *Rendic. R. Accad. Sci. Fis. Mat. Napoli* **16**: 29 (1877).

Type: *Beccari s.n.* Ramoi New Guinea. (Holotype FL).

Dist.: From Ceram, throughout New Guinea to Rossell Is.

Ecol.: Terrestrial in mature rain forest from sea-level - 1500 m. Locally frequent. On uplifted coral.

Specimens: ANDAI: 2.5 km W of Andai - *Johns* 8106. WITHOUT LOCALITY: *Beccari* 12357; *Beccari* 12358; *Sands* 6231.

Notes: Rachis to 90 cm. Fronds with c. 22 pinnae and 3 apical pinnae. Largest pinna to 20-25 cm long.

Tapeinidium moluccanum

Tapeinidium amboynense (Hooker) C.Chr.

Specimens: *Beccari* 12339.

Notes: There are several collections from Sorong (van Royen, Pleyte).

Tapeinidium novoguineense Kramer

Dist.: Throughout New Guinea to Solomon Islands.

Ecol.: On forest floor on steep ridges in lower montane forest. 625 m alt.

Specimens: ARFAK PLAINS: Steep ridges to G. Itsiwei - *Sands* 6367. .

Notes: Plants erect to 60 cm tall. Stipes to 30 cm, dark chocolate-brown. 8 large pinnae and apical section.

LOMARIOPSIDACEAE

Holttum, R.E. (1978). *Lomariopsis* Group. *Flora Malesiana* Ser. II Vol. **1(4)**: 255-330
Holttum, R.E. (1966). *Blumea* **14**: 224

BOLBITIS Schott

***Bolbitis* sp.**

1. **Ecol.**: On coralling limestone in swampy groun near stream in shade at 30 m.
Specimens: KAMPONG MANDOPI: On disturbed hillside in swampy area - *Sands* 6466.
Notes: Erect fern to 1 m tall. Stipe to 30 cm. 14 pairs of pinnae.

ELAPHOGLOSSUM J.Sm.

Elaphoglossum annamense C.Chr. & Tard.
Dist.: Borneo and New Guinea.
Ecol.: In monatne forest. 1900 m altitude.
Specimens: ANGGI LAKES: Anggi Gita Lake - *Kostermans* 2412 (L).
Notes: Known from a single collection from Anggi Lakes.

Elaphoglossum archboldii Copel. *Univ. Cal. Publ. Bot.* **18**: 226 (1942). Copeland (1949: 404); Holttum (1978: 299). **Type**: *L.J. Brass* 13220. Irian Jaya (Holotype GH).
Dist.: Endemic in Irian Jaya.
Ecol.: Epiphyte in moss cushions from 850-2000 m alttiude.
Specimens: KEBAR VALLEY: Mt Nettoti - *van Royen & Sleumer* 8150 (L).
Notes: Fronds fleshy. Sporangia green, black when mature.

Elaphoglossum callifolium (Blume) T.Moore
Dist.: Throughout Malesia.
Ecol.: Epiphyte in montane forest, from 250-2000 m altitude.
Specimens: ARFAK MTS.: Mt Antop near Minjambau - *Chr. Versteegh* BW 12609 B (L).
Notes: Common
Hata L. = 'Ngejoi'.

Elaphoglossum heterolepium Alderw. *Bull. Jard. Bot. Btzg.* II, **16**: 13 (1914). Alderw. (1917: 426). **Type**: *Rachmat* 522. Central Celebes (BO).
Dist.: N Borneo and Celebes to New Guinea.
Ecol.: Epiphyte and lithophyte at 1000-2300 m altitude.
Specimens: KEBAR VALLEY: S slopes of Mt Nettoti along path from Andjai to Saokorem - *van Royen & Sleumer* 3927 (L).
Notes: The type collection is mixed.

Elaphoglossum spp.

1. **Ecol.**: Epiphyte in ridge top lower montane forest dominated by *Araucaria*, *Lithocarpus* and *Litsea* at 650 m altitude.
   Specimens: ARFAK MTS.: Mokwam Trail - *Johns* 7801.
   Notes: Rhizome short-creeping. Stipe to 7-10 cm long, densely covered with broad, brown scales. Fronds dimorphic. Sterile fronds to 26 x 5.5 cms, fertile fronds to 16 x 3.5-4 cms.

2. **Specimens**: ARFAK MTS.: Mokwam Trail - *Croxall* 17..

3. **Ecol.**: Epiphyte on tree boles in subalpine forest to west of summit area at 2000 m altitude.
   Specimens: KOEBRE MTS.: Along track from Iray to summit - *Johns* 8071.
   Notes: Rhizome long-creeping. Fronds dimorphic. Sterile fronds, stipe to 11 cm, fronds to 17 x 2.2 cms. Fertile fronds, stipe to 17 cm, fronds to 16.5 x 2.4 cms.

LOMAGRAMMA J.Sm. in Hooker

Lomagramma prob. ***leucolepis*** Holttum *Blumea* **14**: 234 (1966). Holttum (1978: 287). **Type**: L.J. Brass 24950. Irian Jaya (L).

Dist.: Irian Jaya.
Ecol.: Climber on small tree in shade in lowland rain forest at 0-20 m alt. Type collected in transitional *Lithocarpus* -rain forest. 900 m. Climber to 2-3 m.
Specimens: ARFAK PLAINS: Wariori Flood Palins - Sands 6185.
Notes: Possibly a new species but needs careful checking.

Lomagramma* cf. *melanolepis Alderw. *Bull. Jard. Bot. Btzg.* III **5**: 212 (1922). Holttum (1937: 208); Holttum (1978: 287). **Type**: *Beguin* 1114. Ternate. (Holotype BO).
Dist.: Moluccas, New Guinea to d'Entrecasteuax Is.
Ecol.: In lowland rain forest to 900 m alt.
Specimens: Sands 6206

LOMARIOPSIS Fee

Lomariopsis intermedia (Copel.) Holttum
Ecol.: Climbing on tree stump in shade in disturbed lowland rain forest. 0-20 m alt.
Specimens: ARFAK PLAINS: Sungai Wariori crossing - *Sands* 6271.
Notes: Climber. Sterile fronds with 8 lobes.

Lomariopsis* c.f. *Lomariopsis kingii
Ecol.: Climbing in shade on steep ridge in montane forest. 800 m alt.
Specimens: KEBAR VALLEY: Bukit Moropim - *Sands* 6488.
Notes: Climber to 2m. 28 fertile pinnae.

Lomariopsis sorbifolia (L.) Fee Christ in Warb. (1900: 56).
Stenochlaena sorbifolia J.Sm. Gepp in Gibbs (1917: 73). [see Holttum (1978: 262)]; Holttum (1978: 262); Lauterbach (1929: 568); Lauterbach (1930: 423); K. Schumann & Lauterbach (1900: 131).
Dist.: Pantropical.
Ecol.: Massed along forest margins. 2100 m.
Specimens: ARFAK MTS.: *Beccari* s.n. (FI). ANGGI LAKES: *Gibbs* 5590 (BM).

TERATOPHYLLUM Mett. *ex* Kuhn

Teratophyllum aculeatum (Bl.) Mett. ex Kuhn. *Ann. Mus. Bot. Lugd.-Bat.* **4**: 296 (1869). Holttum (1932: 284); Holttum (1978: 269);
Lomaria aculeata Bl. *Enum. Pl. Jav.* 205 (1828). **Type**: *Blume s.n.* W. Java (Holotype L).
var. aculeatum
Dist.: Throughout Malesia.
Ecol.: In moist areas in lowland rain forest. Young plants growing on exposed tree roots and on bases of tree trunks. Climbing to 5 m or more.
Specimens: Sands 6201
Notes: Sands 6201 is a collection of the distinct bathophylls which occur in this species.

Teratophyllum **sp.**
1. **Ecol.**: Climber in disturbed lowland rain forest on Wariori flood palin at c. 50 m altitude.
Specimens: ARFAK PLAINS: Wariori River- *Johns* 8130.

Notes: Fronds in 3's along rhizome. Fronds strongly dimorphic. Stipe distinctly articulate to rhizome. Stipe swollen at base, green. Sterile fronds, pinnae to 20 x 2.7 cms. Ferile fronds arising in 4's from rhizome, pinnae to 16-17 cms long, very narrow.

***Teratophyllum* spp.**
Specimens: ARFAK PLAINS: Wariori River- *Johns* 8143. Masni - *Johns* 8164. KEBAR VALLEY: *Johns* 8339; *Johns* 8148.

THYSANOSORIA Gepp. in Gibbs

Thysanosoria pteridiformis (Cesati) C.Chr. *Ind. Fil. Suppl.* **3**: 187 (1934). Pichi Sermolli (1977: 91); Holttum (1978: 263); Johns (1996:).
Gymnogramme pteridiformis Cesati *Rend. Acad. Napoli* **16**: 30 (1877). **Type**: *Beccari s.n.* Andai. (Holotype Fl. Herb. Beccari 12704: K).
Thysanosoria dimorphophylla Gepp in Gibbs *Dutch NW New Guinea* 193, pl. 4 (1917). **Type**: *L.S. Gibbs* 6162. (Holotype BM).
Dist.: Kepala Burung. Arfak Plains to Manokwari.
Ecol.: Coral limestone forest near the coast in Manokwari and in alluvial forest on Arfak Plains and foothill forest to S of plains. Abundant in foothill forest to S of Arfak Plains. 0-70 m altitude.
Specimens: ARFAK PLAINS: Wariori River Floodplains - *Johns* 8131; Sands 6284. Foothill Forest S of Wariori floodplain - *Johns* 8249. MANOKWARI: *Beccari* 12704 (Type FL.); Gibbs 6162 (BM).
Notes: Climber with long-creeping rhizome on tree trunks in forest in partial shade. Fronds strongly dimorphic. See Table with collections *Johns* 8131 and *Johns* 8249.

MARATTIACEAE

Angiopteridaceae
Christenseniaceae

Holttum, R.E. (1976). The morphology and taxonomy of *Angiopteris* (Marattiaceae) with description of a new species. *Kew Bull.* **32**: 587-594.

Johns, R.J. (1981). The ferns and fern-allies of Papua New Guinea. Part 7: The Marattiaceae (in part). P.N.G. Univ. Tech. Research Rep. R 48-81: 7.1-7.16.

ANGIOPTERIS Hoffman

van Alderwerelt van Rosenburg, C.R.W.K. (1918). Two critical fern genera: *Angiopteris* Hoffman and *Lecanopteris* Blume. *Bull. Jard. Bot. Btzg.* ser. 2, **28**: 57 - 64.

Holttum, R.E. (1978). The morphology and taxonomy of *Angiopteris* (Marattiaceae) with description of new species. *Kew Bull.* **32**: 587-594.

Pichi-Sermolli, R.E.G. (1957). Notes on nomenclature for genera of Angiopteridaceae. *Webbia* **12**: 339 - 343.

According to Christ the genus is monotypic, but some authorities recognise up to 200 species. As noted by Holttum (1978) the genus requires detailed field studies as many important taxonomic characters are lost when the specimens are dried.

Angiopteris evecta (Forst. f.) Hoffm. *Comment. Soc. Reg. Gotting.* **12**: 29. t.5 (1876). Beccari (1886: 52 as `erecta'); Schlechter in Brause (1912: 3); Cessati (1877: 27); Gepp in Gibbs

(1917: 196); Mueller (1885a: 21); Mueller (1885b: 168); Mueller (1887: 45); Ridley (1916: 264); Schumann & Lauterbach (1900: 146); Wagner and Grether (1948: 26); Christ in Warburg (1900; 94).

Dist.: Palaeotrop. Tropical Africa, Madagascar, Tropical Asia, Indo-Malaya, India, Irian Jaya (Arfak Mts, SW), NE SE New Guinea Bismarck Archipelago, Admiralty Is. Australia, Tahiti.

Specimens: ARFAK MTS.: Putat - *Beccari* s.n. (FI).

Notes: Hieron (1920) lists *A. hellwigii* as a synonym of this species.

Kalkman BW 6311 from Wersar (6km S of Teminaboean) is collected from the Sorong Kabupaten, just to the West on the N coast of the bay.

Angiopteris sp.

1. **Ecol.**: Aifat River Valley in riverine forest near water at 540 m altitude.

Specimens: KEBAR VALLEY: Aifat Valley - van Royen and Sleumer 6881.

Notes: This collection is very distinct from A. evecta. The fronds are quite narrow to 1 cm broad The stem at base is c. 5 cm borad, the plant 5-6 m high.

CHRISTENSENIA Maxon

Ching, R.C. (1940). Christenseniaceae. A new fern family in tropical Asia. *Bull. Fan Mem. Inst. Biol.Bot.* **10**: 227-229.

Christensenia aesculifolia (Bl.) Maxon

Dist.: A single variable species from Assam to the Solomon Islands

Ecol.: In hill forest dominated by *Lithocarpus* and *Hopea*. Along valley floor on earth bank near stream. In shade. 550 m altitude.

Specimens: ARFAK PLAINS: Foothills along S margin of Plains - *Sands* 6319. On steep rdiges - *Sands* 6434.

Notes: This is the first record of this species from mainland New Guinea. See Camus in Kramer and Green (1990: 178).

MARATTIA Swartz

Marattia andaiensis Alderw. *Bull. Btzg.* **23**: 16 (1916). Alderw. (1916: 449). **Type**: Teysmann s.n. Andai, New Guinea. (BO).

Dist.: New Guinea.

Ecol.:

Specimens:

Notes: Closely related to ***Marattia silvatica*** Bl. but the sori are smaller and the margins are lobed, not serrate.

Pinnae shorter towards base than apex. Synangia (8-) 9- to 12-slits.

Marattia fraxinea Smith **Type**: ex Bourbon.

Dist.: Tropical Africa and Asia, throughout Malesia (Java, Moluccas, New Guinea) to Australia.

Ecol.: 100 m.

Specimens: WAREN: 60 km S of Manokwari - Kanehira & Hatusima 14198 (BO).

Notes: This species has large fronds with pinnae up to 19 x 4 cm.

Marattia sambucina Bl. *Rac. Flor. Buitenz.* I: 2. Beccari (1886: 52); Cesati (1877:); Vriese, de (: 6). **Type**: Blume (Holotype L).

Dist.: Malaya, Sumatra, Java, Borneo, Philippines, Flores, Celebes, Moluccas, Irian Jaya.

Ecol.: (450-) 920-2000 m altitude. In Irian Jaya - Arfak Mts to Nassau Geb.

Specimens: ARFAK MTS.: Putat - Beccari s.n.

Notes: Synangia 4- to 8-celled.

Marattia sambucina Bl. is listed as a synonym of this species. However synangia is 12- to 15-celled.

Marattia werneri Rosenstock *Rep. Sp. Nov.* 5: 44 (1908). Alderw. 91908: 767); Brause (1920: 220); Schltr. in Brause (1912: 4). **Type**: Werner 78. (Holotype: B).

Dist: New Guinea. Endemic.

Ecol.: Common terrestrial along valley sides and valley floor. In lower montane forest dominated by *Araucaria*, *Lithocarpus* and *Litsea*, and mid-montane forest on forest floor in shade. Also in mist forest. 650-1900 m altitude.

Specimens: ARFAK PLAINS: On steep ridges to G. Itiwei - Sands 6416. ARFAK MOUNTAINS: Mokwam Trail - *Johns* 7783. ANGGI LAKES: Kanehira & Hatusima 13742. WITHOUT LOCALITY: *van Royen & Sleumer* 7255.

Notes: Stipule to 9x4-5 cm.

Fronds to 3 m long with stipe to 1 m. Pinnae in 15-20 pairs.

Synangia 9-11-celled.

Marattia* sp.** (not *werneri*) = ***Angiopteris

Ecol.: In riverine forest in deep shade in lower montane zone. 450 m alttiude.

Specimens: ARFAK MTS.: Mupi Dessa Trail to G. Humibou - *Sands* 6726.

Notes: Large spreading fern with fronds to 6-7 m long, stipe to 3m long. Fronds with 20-25 pairs of pinnae each with 40-50 pinnules.

***Marattia* spp.**

1. **Ecol.**: Common terrestrial along valley floor and lower valley sides in mature rain forest on uplifted coral.

   Specimens: **ANDAI**: 2.5 km W of Andai - *Johns* 8107.

   Notes: Rachis dotted with blunt 'thorns'. Fronds to 2 m long with c.10 pinnae, to 80 cm long. Margins of pinnae distinctly serrate. Sori submarginal.

2. **Specimens**: WITHOUT LOCALITY: *van Royen & Sleumer* 8096.

MATONIACEAE

Johns, R.J. & Bellamy, A. (1979). The ferns & fern-allies of Papua New Guinea. Pt. 3. The Matoniaceae. pp. 17.1-17.7. P.N.G. Office of Forests.

Kato, M. (1993). A taxonomic study of the genus *Matonia* (Matoniaceae). *Blumea* **38**(1): 167-172.

MATONIA Copel.

Matonia foxworthyi Copel. *Philip. J. Sci.* **3**: 344. t.3 *Bot.* (1909). Alderw. (1916: 114); C.Chr. & Holttum (1934: 223); Copeland (1917: 388); Iwatsuki & Kato (1984: 61); Kato (1993:

171); Tan & Tolentino (1987: 536). **Type**: Foxworthy 372. Mt Poe (Rumput), Sarawak, Borneo. 1700 m alt. (Holo. destroyed; Iso MICH).

Ecol.: In clearings and along margins of moss forest on ridges and summit areas at (429-)750-2100 m altitudes.

Dist.: Borneo, Philippines, Moluccas, to New Guinea. Irian Jaya.

Specimens: ARFAK MT. : Kanehira & Hatusima 14039 (BO); VOGELKOP PENINSULA: van Royen & Sleumer 7204 (L).

NEPHROLEPIDACEAE

NEPHROLEPIS Schott

Nephrolepis acuminata Kuhn Gibbs (1917: 72);

Dist.: Malesia. In New Guinea Arfak Mts., Mt Jaya.

Ecol.: Climbing in forest. 2100 m.

Specimens: ANGGI LAKES: *Gibbs* 6123 (BM).

Notes:

Nephrolepis acuta

Specimens: *Beccari s.n.* (FI).

Nephrolepis biserrata

Specimens: *van Royen & Sleumer* 5035.

Nephrolepis hirsutula (G. Forst.) C. Presl

Ecol.: In warm soils near hot springs. 625 m altitude and on old gardens sites on limestone escarpment at 800 m altitude.

Specimens: KEBAR VALLEY: Near outflow from hot springs - *Sands* 6629. Apni Valley, E of Andjai - *Johns* 8281. ARFAK MTS.: Mokwam Trail - *Sands* 6134. WITHOUT LOCALITY: *van Royen & Sleumer* 3819 (L); *Sands* 6687.

Notes: Plants to 60 cm.

Nephrolepis sp. nov. aff. ***hisutula***

Ecol.: In ridge forest on track from Warmare to Mokwam at 800-850 m.

Specimens: ARFAK MTS.: Mokwam Trail - *Johns* 7705.

Notes: Epiphyte on tree trunk at 2 m.
Non hirsute but otherwise similar to *N. hirsutula.*

***Nephrolepis* spp.**

1 **Ecol.**: In ridge forest on track from Wamare to Mokwam at 800-850 m altitude.
Specimens: ARFAK MTS.: Mokwam Trail - *Johns* 7685.
Notes: Fronds pinnate with 60+ pinnae. Bottom 8-10 pinnae infertile. Veins free. Sori with a distinct green indusium.

2. **Ecol.**: Terrestrial in small gaps in ridge forest along track from Warmare to Mokwam at 800-850 m altitude.
Specimens: ARFAK MTS.: Mokwam Trail - Johns 7717.
Notes: Stipes to 32 cm long, reddish black. Pinnate. Pinnae sessile with largest pinna 6-10th from base.

3. **Specimens**: ANGGI LAKES: Iray - *Johns* 8094.

4. **Specimens**: KEBAR VALLEY: Hot Springs - *Johns* 8373.

5. **Specimens**: AREA 18: *Johns* 8386.

OLEANDRACEAE

Copeland, E.B. (1940). Oleandrid ferns (Davalliaceae) of New Guinea. *Philipp. J. Sci.* **73**(3): 345-357.

Holttum, R.E. (1966). The genus *Arthropteris* J.Sm. in Malesia. *Blumea* **14**: 225-229.

ARTHROPTERIS J.Smith in Hooker f.

Arthropteris articulata (Brack.) C.Chr.
Dist.: Celebes and Moluccas to New Guinea.
Ecol.: Epiphytic climber in *Lithocarpus* forest.
Specimens: KEBAR VALLEY: Path from Son Village to Aifatkanna - *van Royen & Sleumer* 7839 (L).

Arthropteris wollastonii (Ridl.) Holttum *Blumea* **14**: 229 (1966).
Polypodium wollastonii Ridley *Trans. Linn. Soc. Bot.* **9**: 262 (1916). **Type**|: C.B. Kloss s.n. Mt carstensz. (K).
Dist.: New Guinea to Philippines.
Ecol.: Climbing on logs in lowland rain forest, to 800 m altitude.
Specimens: ARFAK MTS.: Mupi Dessa - *Sands* 6834.
Notes: First record from Irian Jaya.

OLEANDRA Cavanilles

Oleandra archboldii Copel. *Philipp. Journ. Sci.* **73**: 346 (1940). **Type**: L.J.Brass 13002. Bernhard Camp, Idenburg River. Irian Jaya. (Holotype Herb. Copeland: BM, BO, K, L).
Dist.: Endemic in New Guinea.
Ecol.: Scandent terrestrial forming shrubs to 1 m tall in rain forest or an epiphytic creeper.
Specimens: MT. NETTOTI: *van Royen & Sleumer* 7485 (L). ANGGI LAKES: Mt Gwamongga - H. Sleumer & W.Vink BW 14245 (L).

Oleandra atricapilla *sp. nov.* manuscript name.
Ecol.: Terrestrial in an open, dry, low, forest.
Specimens: ANGGI LAKES: Kanehira & Hatusima 13509 (B).

Oleandra crassipes Copel. *Philipp. Journ. Sci.* **73**: 347 (1940). **Type**: L.J.Brass 12109. Bernhard Camp, Idenburg River. Irian Jaya. (BM, L, K).
Dist.: Endemic in New Guinea.
Ecol.: Epiphytic or terrestrial. A widespread species of montane forest, particularly *Castanopsis* forest.
Specimens: MT. NETTOTI: *van Royen & Sleumer* 3870; *van Royen & Sleumer* 7373; BW 10420.

Oleandra cuspidata Baker **nom. nov.** Gibbs (1917: 72).

Oleandra articulata Cesati *Felci. Becc. Polin.* 4. ***non*** Cav.

Dist.: Moluccas to New Guinea. Widespread.

Ecol.: Growing in clumps around edge of montane and subalpine forest. 2500 - 3000 m altitude. Sometimes scarse. Forming a shrub to 1-2 m tall.

Specimens: ARFAK MTS. : *Beccari* 12588 (FI); *Beccari s.n.* (FI). ANGGI LAKES: *Gibbs* 5559 (BM); Iray - *Johns* 8000; Summit of Mt Mesenuk - *Sleumer & Vink* BW 14084 (L). Mt Gwamongga - *Sleumer & Vink* BE 14245 (L). KEBAR VALLEY: Along track from Andjai to Saokore -*van Royen & Sleumer* 3842; Mt Nettoti - BW 10299. WIJE VALLEY: *van Royen & Sleumer* 7651 (L).

Notes: Woody shrub to 2m tall.

Oleandra neriiformis Cav.

Specimens: WITHOUT LOCALITY: BW 10419.

Oleandra werneri Ros. *Fedde Repert.* **5**: 40 (1908). **Type**: E. Werner 12. Gelu Mass. New Guinea. (OXF, K, L).

Dist.: Ceram and Moluccas. Widespread throughout New Guinea and Solomon Islands. 800-1300 m altitude.

Ecol.: Common pendulous or scrambling epiphyte in oak forest. A large, shrubby, scrambling epiphyte in mossy forest. 1150 m altitude.

Specimens: NETTOTI MTS.: Chr. Versteegh BW 10424.

Notes: Fronds strongly dimorphic.

Oleandra **spp.**

1. **Ecol.**: In forest floor on lower montane forest at 900 m.
   Specimens: ARFAK PLAINS: On ridge to G. Itsiwei - *Sands* 6399.
   Notes: Terretrial woody fern to 1 m in height.

2. **Ecol.**: Scrambling epiphyte in forest to west of summit at 2000 m altitude.
   Specimens: ANGGI LAKES: Iray - *Johns* 7986.
   Notes: Phyllopodia to 3 cm long.

3. **Ecol.**: In subalpine forest to west of summit at 2000 m altitude.
   Specimens: ANGGI LAKES: Iray - *Johns* 8069.

3. **Ecol.**: In subalpine forest to west of summit at 2000 m altitude.
   Specimens: ANGGI LAKES: Iray - *Johns* 8070.

OPHIOGLOSSACEAE

Jermy, A.C. & Walker, T.G. (1977). A note on the cytology of *Botrychium lanuginosum* and the occurence of the genus in Malesia. *Gard. Bull. Sing.* **30**: 293-298.

Johns, R.J. (1981). The ferns and fern-allies of Papua New Guinea. Part 6. The Ophioglossaceae. *P.N.G. Univ. Tech. Res. Rep.* **R 48-81**: 6.1-6.26.

St. John, H. (1943). Distribution of *Ophioglossum* on the islands of the Pacific Ocean. *Occ. Pap. Bish. Mus.* **17**: 177-182.

Weiffering, J.H. (1964). A preliminary revision of the Indo-Pacific species of *Ophioglossum* (Ophioglossaceae). *Blumea* **12**: 321-337.

HELMINTHOSTACHYS Kaulf.

Helminththostachys zeylanica (L.) Hook. *Gen. Fil.* (1840) t.47. Beccari in d'Albertis (1880) 400; Beccari (1886) 53; Brause (1920) 221; Christ in Warburg (1900) 94; Copeland (1909) 8. t.3; Copeland (1911) 67; Copeland in C.T. White (1929) 272; Dukkus (1930) 156; Kuhn in K.Sch. & Hollr. (1889) 10; Lam (1945) 56; Mueller (1876) 75; Schltr. in Brause (1912) 2; Wagner & Grether (1948) 24); Johns
Osmunda zeylanica L. *Sp. Pl.* (1753). **Type**:

Dist.: Sri Lanka, India, Japan, Formosa, Java, Philippines to New Caledonia and N Australia. Throughout Papuasia: NW NE SE New Guinea to New Ireland and Admiralty Islands.

Ecol.: Terrestrial in swamp forest. In alang fields. Along creekbanks near Prauwen-bivak. 75 m altitude.

Specimens: GUNUNG BORAI: *Sands* 6176. KEBAR VALLEY: Kampong Akmuri - *Johns* 8325. WITHOUT LOCALITY: *van Royen & Sleumer* 5053.

Notes:

OPHIOGLOSSUM L.

Ophioglossum cumingianum Presl Beccari (1886: 53); Cesati (1877: 28).

Distribution: W New Guinea

Ophioglossum petiolatum (Clausen 1938).
Ophioglossum reticulatum (Wief. 1964).

Specimens: Beccari s.n.

Ophioglossum gramineum Willd. *Nov. Act. Acad. Erf.* 2 (1802) 18. t.1. f.1. Beccari (1886) 53; Brause (1920) 220; Cesati (1877) 28; Clausen (1938: 161); Copeland (1909: 4).

Dist.: Africa, India, W New Guinea

Ecol.: In disturbed grassland.

Specimens: ARFAK MTS.: *van Royen & Sleumer* 8254 (L).

Ophioglossum pendulum L. in Stickm. *Herb. Amb.: 27* (1754). Anon. (1899) 121; Baker (1886) 53, 111; Brause (1920) 221; Cesati (1877) 28; Copeland (1909) 5; Lam (1924) 5; Mueller (1885a) 21. pl.2; Mueller (1885b) 168; Ridley (1916) 264; K.Sch. & Laut. (1900: 147). **Type**:

Dist.: Madagascar, Mauritius, Mascarenes, SE Asia, Sri Lanka, India, Formosa, Japan (Yakushima), throughout Malesia to New Caledonia, Polynesia, Sandwich Islands, Hawaii and NSW Australia. In Papuasia: NW SW NE SE New Guinea, Admiralty Islands, New Ireland.

Specimens: *van Royen & Sleumer* 7892 (L).

Ophioglossum **spp.**

1. **Dist.**: Along steep ridges between Arfak Plains and Gunung Itsiwei.
   Ecol.: Terrestrial in shade in ridge forest dominated by *Lithocarpus* and *Shorea*.
   Specimens: ARFAK PLAINS: Mountains South - *Sands* 6418.

2. **Specimens**: ARFAK MTS.: Mupi Valley - Mahyar 329.

OSMUNDACEAE

Hennipman,E. (1968). A new *Todea* from New Guinea with remarks on the generic delimitation in recent Osmundaceae. *Blumea* **16**: 105-108.

TODEA Willdenow in Bernhardi

Todea barbara (L.) Moore
Dist.: South Africa, Australia, New Zealand.
Specimens: ARFAK MT.: Hatam - Beccari s.n. (FI). KEBAR VALLEY:
Notes: The New Guinea species has been described as *Todea papuana* Hennipman.

PARKERIACEAE

Johns, R.J. and Bellamy, A. (1979). The Ferns and Fern-Allies of Papua New Guinea. Part 2. The Parkeriaceae. P.N.G. Office of Forests 11.1-11.5.

Lloyd, R.M. (1972). Species delimitation in the genus *Ceratopteris* (Parkeriaceae). *Amer. J. Bot.* **59**: 676.

CERATOPTERIS Brongn.

Ceratopteris thalictroides (L.) Brongn. Bull. Soc. Philom. 186 (18). Ceasati (1877: 25); Christ. in Warburg (1900: 93); Copeland (1947: 83); Muell. (1885: 21); Schumann & Lauterbach (1900: 144).

Acrostichum thalictroides L.

Dist.: Widely distributed throughout Asian Tropics to New Guinea..
Ecol.: Floating aquatic, often at water's edge. Sea-level to 1200 m alttiude. Grows in various wet and flooded sites.
Specimens: *Sands* 6211.
Notes: Plants reportedly cultivated for food in some areas in New Ireland.

PLAGIOGYRIACEAE

PLAGIOGYRIA

Plagiogyria sp. near ***Plagiogyria tuberculata*** Copel. Brass (1941: 298,314); C.Chr. (1937: 284); Posth. in van Steenis (1934: 173).
Dist.: Sumatra, Borneo, Philippines to New Guinea.
Specimens: MT KOEBRE: *Johns* 8081.

POLYPODIACEAE

Polypodiaceae sp.

1. **Ecol.**: In wet areas and swamps to south of range at 900 m altitude.

Specimens: KEBAR VALLEY: Apni Valley - *Johns* 8284.
Notes: Rhizome mid-creeping. Stipe to 9 cm long. Fronds simple.

2. Specimens: ARFAK MTS.: Mokwam Trail - *Johns* 7791.

3. Specimens: ARFAK PLAINS.: Wariori Camp - *Johns* 8201.

BELVISIA Mirbel

Belvisia mucronata (Fee) Copel.
Ecol.: Epiphyte on fallen trunk in lower montane forest at 700 - 900 m altitude.
Specimens: ARFAK PLAINS: On steep ridge to G. Itsiwei - *Sands* 6402. ARFAK MTS.: Sungai Mupi - *Sands* 6824. KEBAR VALLEY: Trail from Andjai to Mt Nettoti - *Davis* 710.
Notes: Epiphyte with sterile fronds to 25 cm and fertile fronds to 45 cm. Terminal soral region to 7 cm.

Belvisia novoguineensis (Rosenst.) Copel.
Specimens: *Mayar* 411.

***Belvisia* spp.**

1. **Ecol.**: Epiphyte on tree trunk over stream in partial shade. Montane forest at 825 m altitude.
   Specimens: ARFAK MTS.: Mupi Dessa Trail - *Sands* 6824.
   Notes: Epiphyte with tufted habit. Stipes brownish-green.

2. **Ecol.**: Epiphyte on tree trunk of *Lithocarpus* in ridge forest in montane forest at 825 m altitude.
   Specimens: ARFAK MTS.: Mokwam Trail - *Johns* 7688.
   Notes: Epiphyte, climbing on tree trunk. Stipes black to 2-3 cm, fronds to 26 cm.

3. **Ecol.**: Epiphyte on tree trunk in subalpine forest at 2500-3000 m altitude.
   Specimens: MT. KOEBRE: Along Trail through forest patches - *Johns* 8055.
   Notes: Epiphyte, rhizome short-creeping. Stipes black to 2.5 cm, fronds to 23 cm long, fleshy. Fertile region to 12 cm long.

4. **Ecol.**: Epiphyte in forest at 1700-1940 m near Iray.
   Specimens: ANGGI LAKES: Iray - *Johns* 7988.
   Notes: Creeping rhizome. Stipe to 6 cm long. Fronds fleshy, to 36 cm with apical fertile region to 10-11 cm.

5. **Specimens**: ANGGI LAKES: Iray - *Johns* 7985.

DRYNARIA (Bory) J.Smith

Drynaria rigidula (Sw.) Bedd.
Dist.: Sri Lanka to Pacific.
Ecol.: Epiphyte on fallen *Hopea* in lower montane forest at 650 m. Sealevel to 1300 m.
Specimens: ARFAK PLAINS: On ridges to G. Itsiwei - *Sands* 6408. KEBAR VALLEY: E of Andjai to Mt Nettoti - *Davis* 829.

Notes: Sheathing fronds brown. Fertile fronds to 70 cm.

Drynaria sparsisora (Desv.) T.Moore
Ecol.: Climber on roots of *Pandanus* along ridgetop on limestone escarpment at 900 m altitude.
Specimens: KEBAR VALLEY: E of Andjai 2 - *Johns* 8278.
Notes: Rhizome long-creeping. Distinct nest and fertile fronds. Ferile fronds to 50 cm long.

***Drynaria* spp.**
1. **Ecol.**: A creeping terrestrial on limestone outcrop in disturbed forest at 600 m altitude.
Specimens: KEBAR VALLEY: On limestone outcrop - *Johns* 8352.
Notes: Nest leaves to 8 x 6 cms on slender rhizome. Stipe to 14 cm, winged. Fronds 4-5 lobed, to 30 cm long.

DRYNARIOPSIS

Drynariopsis heraclea (Kunze) Ching
Aglaomorpha heraclea (Kunze) Copel.
Ecol.: Epiphyte in lowland forest in shade to 760 m altitude.
Dist.: Malesia to Solomons.
Specimens: ARFAK MTS.: Mupi Dessa - *Sands* s.n.

GONIOPHLEBIUM

Goniophlebium percussum (Cav.) W.H. Wagner & Grethet
Goniophlebium verrucosum Wal. ex Hook. *Garden Ferns* t. 41 (1862).
Ecol.: Hill to montane forest at 800 m.
Specimens: BUKIT MOROPIM: *Sands* 6507.

LECANOPTERIS Reinwardt

Lecanopteris deparioides (Ces.) Baker
Ecol.: Epiphyte in lower montane forest on trunk of *Lithocarpus* at 2 m height.
Specimens: ARFAK PLAINS: On steep ridge to G. Itsiwei - *Sands* 6378.
Notes: Rhizomes swollen, forming ant houses, glaucous. 28 pinnae.

LEMMAPHYLLUM

Lemmaphyllum accedens (Blume) Donk
Dist.: Malesia to W Pacific.
Ecol.: Epiphytic on branch of low Urticaceous shrub in ridge top lowe montane forest dominated by *Araucaria*, mixed with *Lithocarpus* and *Litsea* at 650 m altitude.
Specimens: ARFAK MTS.: Mokwam Trail: *Johns* 7779. Mupi Dessa Trail - *Sands* 6807.
Notes: Rhizome long-creeping. Fronds to 5x1.3 cm. Stipe to 5 mm. Sori at tips of fronds - 4-20 in each apical region.

MERINTHOSORUS

Merinthosorus drynarioides (Hook.) Copel.
Ecol.: High epiphyte in lowland rain forest at 50 m altitude.
Specimens: ARFAK PLAINS: Masni on Wariori flood plains - *Johns* 8181.

MICROSORUM Link

Microsorum commutatum (Blume) Copel.
Ecol.: Epiphyte in shade in lower montane forest at 810 m.
Specimens: ARFAK MTS. : Mupi Dessa Trail - *Sands* 6836.
Notes: Rhizome dark green. Stipe light green on younger fronds. Fronds caudate.

Microsorum liguiforme (Mett.) Copel.
Dist.: S India to Fiji (Excl Java and Lesser Sundas).
Ecol.: Epiphyte on fallen trunk in partially disturbed lowland and lower montane forest to 900 m altitude.
Specimens: *Sands* 6243. ARFAK PLAINS: On steep ridges to G. Itsiwei - *Sands* 6402A.
Notes: Growth tufted with sterile fronds to 25 cm and fertile fronds to 45 cms.

Microsorum **c.f.** *Microsorum papuanum* (Baker) Parris
Ecol.: Epiphyte in forest patches near Lake Giggi at 900 - 1940 m altitude.
Specimens: ANGGI LAKES: Near Lake Anggi Giggi - *Johns* 7967. ARFAK MTS. : Mupi Dessa Trail - *Sands* 6826.
Notes: Rhizomes long-creeping, bluish-grey. Fronds simple, stipe to 5 cm, reddish-brown. Frond to 14 cm, golden brown sori, prominent above. Fronds to 1 m, coriaceous.
In ***Phymatosorus*** according to Bosman.

Microsorum scolopendria (N.L. Burm.) Copel.
Phymatosorus scolopendria (N.L.Burm.) Pichi Serm.
Ecol.: Climber on lower tree branches in forest at low altitudes.
Specimens: RANSIKI RD.: Johns 7837.
Notes: Rhizome long-creeping, grey-green, covered with dark scaleswith grey margins. Stipe reddish-green to 15-20 cm long. Fronds with 3 lobes and a larger terminal lobe. Largest lateral lobe to 16-17 x 3.5 cm, terminal lobe to 18x4.5 cm.

***Microsorum* spp.**

1. **Specimens**: Sands 6133.

2. **Ecol.**: Epiphyte in shade in partially disturbed lowland rain forest.
Specimens: ARFAK PLAINS: S. Wariori River crossing - *Sands* 6266.
Notes: Epiphyte with long climbing rhizomes. Fronds to 50 cm long, stipes short, to 4 mm.

3. **Ecol.**: In lowland rain forest at 30-50 m altitude.
Specimens: ARFAK PLAINS: Foothills to south of plains - *Sands* 6297.
Notes: Fronds to 60 cm long with base of stipe covered with dark brown hairs.

4. **Ecol.**: Epiphytic climber in shade in lowland rain forest. 30-50 m altitude.
Specimens: ARFAK PLAINS: In foothills to south - *Sands* 6322.

Notes: Rhizome flattened on ventral side. Fronds to 66 cm long, spaced at 7-9 cm along rhizome. Pinnae to 25 cm long. Sori in secondary areolae.

5. **Ecol.**: In forest at 1700-1940 m.
Specimens: IRAY: Johns 7995.
Notes: Rhizome long-creeping, forming a nest of roots. Rhizome whitish-grey with scattered dark black scales. Stipe black. Fronds simple to 1.4 m long.

6. **Ecol.**: In forest on uplifted coral platform at 250 m.
Specimens: IRAY: Johns 7665.
Notes: Rhizome mid- to long-creeping. Stipe apparently adnate to rhizome. Stipe to 58 cm. Pinnatifid with 12-14 pairs of pinnae. Veins forming two distinct sets of aereolae.

7. **Ecol.**: Terrestrial at 650 m altitude.
Specimens: KEBAR VALLEY: S of Andjai 2 - *Johns* 8316.
Notes: Rhizome creeping with roots only on ventral side. Fronds pinnatifid, very fleshy, larger fronds with 6-7 lobes. Stipes to 48 cm long, winged at top. Soral areas depressed below.

PHYMATODES C. Presl

***Phymatodes* spp.**

1. **Ecol.**: Epiphytic in ridge top lower montane forest in *Araucaria, Lithocarpus* and *Litsea* at 650 m altitude.
Specimens: ARFAK MTS.: Mokwam Trail - Johns 7782.
Notes: Rhizome covered with short peltate scales. Stipes to 5 cm. Fertile fronds to 28 x 4 cms. Sterile fronds shorter. Sori protruded above.

2. **Ecol.**: Epiphytic in ridge top lower montane forest in *Araucaria, Lithocarpus* and *Litsea* at 650 m altitude.
Specimens: ARFAK MTS.: Mokwam Trail - Johns 7797.
Notes: Rhizome climbing. Stipe with scattered scales. Fronds pinnatifid with 3-8 lobes.

3. **Ecol.**: Epiphytic along patchily waterlogged lake margin strewn with boulders.
Specimens: ANGGI LAKES: Between end of airstrip and lake - Johns 7921.
Notes: Rhizome long-creeping. Stipe 5-6 cm long, black with a very narrow wing. Pinnae to 26.5 x 2.5 cms.

PHYMATOPTERIS

Phymatopteris albidosquamata (Blume) Pichi Serm.
Ecol.: In montane forest at 1700 - 1940 m altitude.
Specimens: IRAY: Johns 7992.
Notes: Rhizome long-creeping, densely black, covered with scales. Stipe dense black to 20 cm long. Fronds with 5-8 pairs pinnae, largest to 16 x 1.2 cm.

Phymatopteris taeniata (Sw.) Pichi Serm.
 Crypsinus taeiatus (Sw.) Copel.
Ecol.: In montane forest at 1700 - 1940 m altitude.
Specimens: IRAY: Johns 7993.

Notes: Rhizome long-creeping, densely black, covered with scales. Stipe pale green to brown to 30 cm long. Fronds with 10 pairs pinnae, largest to 16 x 1.2 cm. Lacking white aerophores along margin.

A very small form with up to 16 narrow pinnae.

PHYMATOSORUS

1. **Ecol.**: Epiphytic on *Dacrydium* in swamp forest near Anggi Gigi at 1950 m altitude.
Specimens: ANGGI LAKES: *Johns* 8095.
Notes: Rhizome long-creeping. Rachis red. Fronds simple, to 14.5 x 5.5 cm.

2. **Ecol.**: Epiphyte on tree branches in lowland rain forest at 50 - 100 m altitude.
Specimens: RANSIKI RD.: Along roadside in forest inland from Ransiki - *Johns* 7899.
Notes: Stipes to 55 cm long, winged. Fronds pinnatifid, rachis to 25 cm long.

PLATYCERIUM Desv.

Platycerium wandae Racib. Brause (1920: 208); Joe (1964: 107).
Ecol.: High epiphyte on tree branches, often with several plants on a single branch at 0-50 m altitude or in *Aaraucaria* plantations at 600 m altitude.
Specimens: RANSIKI RD.: Along road in forest - *Johns* 7834. KEBAR VALLEY: In *Araucaria* plantation - *Johns* 8326.
Notes: Basal frons broad, detritus collecting. Fronds dimorphic. Fertile fronds with 2-lobes. Apex fertile. Sporangia acrostichoids, paraphyses stellate.

PLEOPELTIS Humb. ex Bonpl. ex Willd.

Pleopeltis argyropus Ridley. Gibbs (1917: 75);
Polypodium argyropus Ridley *Trans. Linn. Soc. Ser. 2 Bot.* **9**: 262 (1916).
Dist.: New Guinea. Arfak Mts to Mt Jaya.
Ecol.: Epiphyte in forest. 2100 - 2550 m.
Specimens: KOEBRE MT. : *Gibbs* 5632 (BM); ANGGI LAKES: *Gibbs* 5589 (BM).

Pleopeltis papuanum (Baker) Gibbs (1917: 75);
Polypodium papuanum Baker
Dist.: New Guinea.
Ecol.: Epiphyte in moss forest. Also terrestrial. 2100 - 2700 m.
Specimens: ARFAK MTS. : *Beccari s.n.* (FI). ANGGI LAKES: *Gibbs* 6010 (BM); *Gibbs* 5933 (BM).

Pleopeltis remigerum Ridley (in sched.) Gepp in Gibbs (1917: 75); Lam (1928:295); Lam (1945: 88).
Polypodium remigerum Ridley
Pleopeltis renifera Ridley *Trans. Linn. Soc. Ser. 2 Bot.* **9**: 263 (1916).
Dist.: New Guinea. Arfak Mts to Mt Jaya.
Ecol.: Epiphyte in moss forest. 2400 - 2700 m.
Specimens: ANGGI LAKES: *Gibbs* 5987 (BM); *Gibbs* 6015 (BM).

Pleopeltis stenophyllum (Bl.) Gibbs (1917: 75);

Polypodium stenophyllum Bl.
Dist.: From Malay Peninsula to New Guinea and Fiji.
Ecol.: Epiphyte in moss forest. 2400 - 2700 m.
Specimens: ANGGI LAKES: *Gibbs* 6015 (BM); *Gibbs* 5987 *pro parte* (BM).

POLYPODIUM L.

Polypodium curtisii Baker
Dist.: Sumatra to New Guinea.
Ecol.: Epiphyte in mossy forest. 2100-2400 m.
Specimens: ANGGI LAKES: *Gibbs* 5664 (BM); *Gibbs* 5733 (BM).

Polypodium hirtellum Bl.
Dist.: South China to New Caledonia.
Ecol.: Epiphyte in forest. 2100 m.
Specimens: ANGGI LAKES: *Gibbs* 5551 (BM);

Polypodium scabristipes Baker
Dist.: New Guinea.
Ecol.: Epiphyte in forest. 2100 m.
Specimens: ANGGI LAKES: *Gibbs* 5972 (BM);

Polypodium serratodentatum Alderw. Gibbs (1917: 75);
Dist.: Java? to New Guinea.
Ecol.: Epiphyte in forest. 2100 - 2400 m.
Specimens: KOEBRE MT. : *Gibbs* 5628 (BM);

Polypodium **spp.**

1. **Ecol.**: Climber to 4 m in deep shade in riverine forest in lower montane zone. 450 m alt.
   Specimens: ARFAK MTS.: Mupi Dessa trail to G. Humibou - Sands 6728.
   Notes: Rhizomes wiry, greyish-green.

2. **Ecol.**: Epiphyte on trunk in subalpine forest at 1950 m altitude.
   Sppecimens: MT KOEBRE: Track form Iray to Summit - *Johns* 8077.
   Notes: Rhizome long-creeping, densley covered with gingery scales. Sterile fronds to 11.5 x 1.4 cm, fertile fronds to 14.5 x 4 cm.

3. **Ecol.**: Epiphyte on trunk in subalpine forest at 1950 m altitude.
   Sppecimens: MT KOEBRE: Track form Iray to Summit - *Johns* 8087.
   Notes: Rhizome long-creeping, densley covered with white scales. Stipe on sterile fronds to 15 cm. Fronds to 11.5 x 1.4 cm, fertile fronds to 14.5 x 4 cm.

PYRROSIA Mirbel

Pyrrosia lanceolata (L.) Farw.
Ecol.: Epiphyte in lowland and hillforest on waeathered limestone. Climbing on trees in shade. 50 - 725 m altitude.
Specimens: BUKIT BASAM: - *Sands* 6601. RANSIKI ROAD: Inland from Ransiki - *Johns* 7892.

Pyrrosia longifolia (N.L.Burm) C.V. Morton
Ecol.: Epiphyte on bamboo in ridge top forest along limestone escarpment at 900 m altitude.
Specimens: **KEBAR VALLEY**: E of Andjai 2 - *Johns* 8276.
Notes: Rhizome mid-creeping, stipes to 3-4 cm, green to black.

Pyrrosia princeps (Mett.) C.V. Morton
Dist.: Sulawesi, Moluccas, New Guinea: widespread.
Ecol.: High epiphyte in lowland rain forest on flood plain at 50 m. Also terrestrial on uplifted coral in mature ridge top rain forest. (Probably fallen from od branch).
Specimens: ANDAI PLAINS: Masni on Wariori flood plain - *Johns* 8182. ANDAI FOREST: 2.5 km West of Andai - *Johns* 8103. KEBAR VALLEY: On limestone outcrop - *Johns* 8357. WITHOUT LOCALITY: *Sands* 6137

Pyrrosia spp.
1. **Ecol.**: Epiphyte in lower montane forest in river valley at 80-150 m altitude.
Specimens: **ARFAK PLAINS**: between Wariori and Mangopi Rivers - *Johns* 8206.
Notes: Fertile fronds to 16 cm, sterile fronds to 9.5 cm. Top of fertile frond slightly contracted.

SELLIGUEA Bory

Selliguea feei (Mett.) Gibbs (1917: 75);
Polypodium feei Mett.
Dist.: Malesia to Polynesia.
Ecol.: Epiphyte and terrestrial near lakes. 2100 m.
Specimens: ANGGI LAKES: *Gibbs* 5948 (BM).

PTERIDACEAE

ACROSTICHUM L.

Acrostichum aureum L.
Dist.: Pantropical..
Ecol.: Terrestrial in swampy areas in disturbed lowland rain forest. 50 m alt.
Specimens: ARFAK PLAINS: c. 5 km inland from coast - *Johns* 8132.
Notes: Fronds to 3 m tall. Sterile pinnae to 35 x 6.5 cm. ower

Acrostichum sp.
1. **Ecol.**: Terrestrial on rocks adjacent to hot steam from hot springs at 600 m altitude.
Specimens: KEBAR VALLEY: Hot springs area S of Andjai 2 - *Johns* 8372.
Notes: Rhizomes forming a distinct caudex. Stipes to 30 cm long. Pinnae alternate, to 19 x 3 cm.

PTERIS L.

Pteris deltoidea Copel.
Ecol.: Lowland forest on steep hillside at 60 m. altitude.
Specimens: RANSIKI ROAD: Between Wanmarwai and Bukit Borai - *Sands* 6701.

Pteris ensiformis Burm. f.
Dist.: India to Pacific including Australasia.
Ecol.: On forest floor in shade in *Araucaria* plantations at 600 m.
Specimens: KEBAR VALLEY: Track to Sungei Aprii - *Sands* s.n.

Pteris ligulata Gaud. *Freycinet. Voyage Bot.* 385 (1827). **Type**: *Gaudichaud*, Rawak, Moluccas (B; K).

Pteris beccariana C.Chr. *Index* 593 (1906). **Type**: *Beccari s.n.* (BM).

Dist.: Moluccas to New Guinea and Solomon Is.
Ecol.:
Specimens:
Notes:

Pteris pacifica Hieron. *Hedwigia* **55**: 355 (1914). **Types**: Nyman 6; Nyman 748; Peekel 41; Schlechter 14524; Warburg s.n. (B).
Notes: Collected from Warsamson Valley, Sorong. No specimens yet seen from Manokwari Kabupaten.

Pteris papuana Ces. *Rend. Ac. Napoli* **16**: 25-29 (1877). Gibbs (1917: 73). **Type**: *Cesati s.n.* New Guinea. (BM; K).

Pteris bambusoides Gepp ex Gibbs *Dutch NW New Guinea* 195 (1917). **Type**: *L.S. Gibbs* 5732. Arfak Mts. Momi River to Wariap. (BM).

Pteris brausei (Rosenst.) Rosenst. in C.Chr. *Ind. Fil. Suppl.* 66 (1913). *Taenitis brausei* Rosenst. *Nova Guinea* **8**: 730 (1912).

Dist.: Widespread in New Guinea, often confused with *P. moluccana*. Also in Solomon Is. and Vanuatu.
Ecol.: Locally comon in grasslands and *Casuarina* forest at low altitudes. Often in disturbed habitats.
Specimens: ARFAK MTS.: *Beccari s.n.* (FI). Momi R. to Wariap - *Gibbs* 5732. ANGGI LAKES: *Gibbs* 5731 (BM). RANSIKI: *Kostermans* 2691 (BO,K). KEBAR VALLEY: Andjai - *van Royen & Sleumer* 6727 (L,K).
Notes: Copeland (1946) considers this synonymous with *P. opaca* J.Sm. a species occuring in the Philippines and the Celebes.

The rhizome is used as a source of flour on Frederik Hendrik Island.

Pteris quadriaurita Retz. Flor. Betz. 11: 156.
Notes: Alderwerelt (1908: 365-366) discussed the many varieties of this species.

Hieronymous (*Hedwigia* **55**: 325-375 (1914)) stated that this species is confined to Sri Lanka.

Pteris rosenstockii C.Chr. *Dansk. Bot. Ark.* **9**(3): 45 (1937). **Type**: G. Bamler 26. 4/1914. Sattleburg. Morobe Province. P.N.G. 900 m. (Holotype: B).
Dist.: New Guinea and the Moluccas.
Ecol.: Locally common in grasslands, along forest margins and on flood banks of rivers. Low to medium altitudes. Locally common. to 550 m.
Specimens: ARFAK MTS.: Herb. *Beccari* 12422 (B). KEBAR VALLEY: *van Royen* 5092; Andjai - van Royen and Sleumer 6727; Versteeg BW 721 (BO).
Notes: Scales on the stipe are long, translucent and slender. The pinnae are serrate for their entire length and have a long attachment to the rachis.

van Royen 5092 as the notation var spinescens nov. var. by Alston.

Pteris semipinnata L. *Sp. Pl.* 1076 (1753).
Dist.: S India and Japan, throughout Malesia. Widespread in New Guinea.
Ecol.: Terrestrial in ridge top forest on limestone.
Specimens: KEBAR VALLEY: E of Andjai - *Johns* 8280.
Notes: Copeland (1946: 33) listed this species from New Guinea: 'it was only listed by report'.

Pteris tripartita Swartz *Schrader's Journal* "1800" 67 (1801).
Ecol.: Disturbed lowland rain forest from 0-20 m altitude.
Specimens: ARFAK PLAINS: Wariori flood plain - Sands 6188. MOETOERI: van Royen 4069 (L).

Pteris vittata L. Sp. Plant. 1074 (1753).
Dist.: A widespread species throughout tropical areas of Asia, Africa and the Pacific. Throughout Malesia very common in New Guinea.
Ecol.: Grows from sea-level to 2000 m. Terrestrial. Often common on soils with calcite. Coastal on uplifted coral, also on sandy banks of streams. In New Guinea from sea-level to 1800 m. Most collections are from low altitudes below 100 m.
Specimens: ARFAK MTS.: *Beccari s.n.* (K). GEELVINK BAY: Versteeg and W. Vink BW 8275. 60 m. (K, L). RANSIKI: 5km NW of Ransiki along road - *Johns* 7876; *Sands* 6688.
Notes: Rhizome erect with dense, small white scales. Stipe to 26 cm long. Rachis to 76 cm, both stipe and rachis grooved. Lower pinnae small to 2 cm long. Largest pinnae to 15x0.9 cm. 40 pairs.

Pteris wallichiana Agardh. *Rec. Sp. Gen. Pteridfis* 69 (1839). **Type**: *Wallich* (K, BM).
Dist.: Throughout New Guinea.
Ecol.: 100-1200(-3470) m alt.
Specimens: ARFAK BEI: *Gjellerup* 1241. (B; L). GEELVINK BAI: Schonian 16.
Notes:

Pteris zippelii

Allosorus zippelii Miquel *Ann. Mus. Lugd. Bat.* **4**: 98 (1868). **Type**: *Zippelius s.n.* New Guinea.
Specimens: *Versteegh & Vink* BW 8275 is from Biak in secondary forest. Should occur in the Vogelkop.
Notes: *Allosorus zippelii* is a juvenile form of *Pteris vittata* L.
BW 8275 is very hairy on the stipe.

***Pteris* spp.**

1. **Ecol.**:
   Specimens: *Johns* 8334.
   Notes:

2. **Ecol.**: Disturbed lowland forest in limestone quarry site and surrounding hills. 30-50 m altitude.
   Specimens: RANSIKI RD. : Bukit Batu Putih - *Sands* 6688.
   Notes: Fronds to 120 cm.

3. **Ecol.**: In lowland forest on steep hill slope in shade. 60 m alt.
   Specimens: RANSIKI RD. : Between Wanmarwai and B. Borai - *Sands* 6701.
   Notes: Fronds to 60 cm, stipe wiry, dark reddish.

4. **Ecol.**: Epiphytic.
Specimens: ANGGI LAKES: *L.S. Gibbs* 5731. 12/1913. 2100 m. (BM; K).
Notes: This species is closely related to *Pteris moluccana* Bl.

5. **Ecol.**: Terrestrial in *Araucaria* forest in shade. 600 m alt.
Specimens: KEBAR VALLEY:
Notes: Erect species with fronds to 70 cm long, stipe blue-green to 30 cm.

6. **Ecol.**: Terrestrial in forest on uplifted coral platform at 250 m altitude.
Specimens: **GUNUNG MEJA**: *Johns* 7675.
Notes: Stipe to 1 m long. Frond margin, black recurved.

7. **Ecol.**: Terrestrial in valley floor forest along track from Wamare to Mokwam at 600-650 m altitude.
Specimens: **ARFAK MTS.**: Mokwam Trail - *Johns* 7754.
Notes: Stipe to 1 m long. Rachis to 1 m long. Veins copiously anastomising.

8. **Ecol.**: Terrestrial in valley floor forest along track from Wamare to Mokwam at 600-650 m altitude.
Specimens: **ARFAK MTS.**: Mokwam Trail - *Johns* 7766.
Notes: Stipe to 63 cm long, base densely covered with large brown scales.

9. **Ecol.**: Terrestrial in ridge top forest on limestone escarpment at 900 m altitude.
Specimens: **KEBAR VALLEY**: E of Andjai 2 - *Johns* 8280.
Notes: Stipes to 60-65 cm long. Fronds pinnate-pinnatifid with 8-10 pairs opposite pinnae. Basal pinnae to 23 cm long with long basal lobes to 15 cm.

10. **Ecol.**: Terrestrial on flood plain in primary forest near Wariori River at 70 m altitude.
Specimens: **ARFAK PLAINS**: Wariori Flood Plain - *Johns* 8227.
Notes: Rhizome long-creeping. Stipe to 50 cm long, rachis to 115 cm with 30 groups and a terminal pinna. Lower pinnae in 3's (9); next 17 in 2's; upper 4 singly.

SCHIZAEACEAE

Holttum, R.E. (1959). Schizaeaceae. *Flora Malesiana* Ser. II. Vol. **1**(1): 37-61.

LYGODIUM Swartz

Lygodium circinnatum (N.L. Burm) Sw.
Dist.: Sri Lanka and India to Malesia and Micronesia.
Ecol.: Climber in disturbed areas. Also in rain forest on dry limestone at 10-260 m altitude.
Specimens: RANSIKI: Along road from Manokwari to Ransiki - *Johns* 7815. Along road 5km NW Ransiki - *Johns* 7882. RANSIKI AREA: 5 km NW of Ransiki - *Johns* 7856, *Johns* 7859. ARFAK PLAINS: Andai Forest Reserve - *Sands* 6887.
Notes: Fronds opposite, 6 lobed. Lobes to 20x2 cm. Fertile fronds terminal on stems, veins simple to 0.6cm wide.

***Lygodium* spp.**

1. **Ecol.**: Climbing in disturbed areas around camp site along track from Wamare to Mokwam at 900-950 m.
Specimens: ARFAK MTS.: Mokwam Trail - *Johns* 7760.
Notes: Apical pinnae fertile, lower pinnae sterile. 5-6 sterile fronds at each node.

SCHIZAEA J.E.Smith

Schizaea dichotoma (L.) Sm. *Mem. Ac. Turin* **5**: 422 (1793). Alderw. (1908: 106); Holttum (1959: 41). **Type**: Petiver. *Gaz.* t.70, f.12 (drawing of a collection from Cochinchina).
Acrostichum dichotomum L. *Sp. Pl.* 1068 (1753).
Dist.: Mascarenes, Ceylon and S India, throughout Malesia to Australia and New Zealand.
Ecol.: Terrestrial in lightly shaded areas in forest, mainly in lower montane forest dominated by *Castanopsis*, *Lithocarpus* and *Agathis*. Sea-level to 1000(-1800) m altitude. Rarely epiphytic.
Specimens: ARFAK PLAINS: Sands 6143. ANGGI LAKES: Bivouac Noordpool, Lake Gita - *Sleumer & Vink* BW 14085 (L). KEBAR VALLEY: G. Sobor - *Chr. Versteegh* BW 714 (L); Mt Sobeto NE of Andjai Village - *van Royen* 4956 (L); Sobor Hills, SE Andjai - *van Royen* 5069 (L).; *Gjellerup* 677 (L). RANSIKI: *Kalkman* 3564 (L). WITHOUT LOCALITY: Sands 6259; Sands 6355. van Royen 3455, 4956, 5420, 6646; van Royen & Sleumer 6450.

Schizaea fistulosa Labill.
Ecol.: In open swampy areas with mosses and *Centrolepis* at 2600 to (3700 m).
Specimens: ARFAK MTS.: Summit area of Mt Sensenemes above Lake Gigi - *Sleumer & Vink* 4339 (L). KEBAR VALLEY: Aifat River Valley - *van Royen & Sleumer* 7291 (L). Mt Nettoti, path to Wekari River - *van Royen & Sleumer* 8059 (L); Mt Nettoti top - *van Royen & Sleumer* 7861 (L); *van Royen* 11065 (L).

Schizaea malaccana Baker *Syn. Fil.* 428 (1868). Gibbs (1917: 76); Holttum (1955: 52); Holttum (1959: 44). **Type**: Cuming 379. Mt Ophir, Malaya (Holotype K: BM).
Schizaea papuana Brause *Bot. Jahrb.* **56**: 211 (1920).
Dist.: Burma to Malesia.
Ecol.: Common on open drier and gravelly parts of marsh near lake. 2100 - 2550 m. Also epiphyte on tree trunks or in moss cushions witn *Xanthomyrtus* and in old secondary forests.
Specimens: ANGGI LAKES: *Gibbs* 5575 (BM); *Gibbs* 6011 (BM). BW 10400; *van Royen* 8024.
Notes: Holttum (1959: 44) recognises two varieties. Both occur in Irian Jaya.

Schizaea malaccana Baker var. ***robustior*** C.Chr. **Type**: Hallier 2150. Sungei Kenepai, Borneo. (L).
Specimens: KEBAR VALLEY: N slopes of Mt Nettoti - *van Royen & Sleumer* 8024 (L); *Chr. Versteegh* BW 10400 (L).
Notes: Herb to 25 cm tall.

Schizaea wagneri Selling *Svensk. Bot. Tidskr.* **40**: 278 (1946). Holttum (1955: 52); Holttum (1959: 44). **Type**: *Grether & Wagner* 4177. Mt Tjajiak. 600 m. Manus Is. (Holotype S-PA; K).
Dist.: Malaya, Singapore, Ambon to Irian Jaya and Manus Is.
Ecol.: Epiphyte in moss on tree trunks. In IJ and Borneo near coast.
Notes: Very inconspicuous with very narrow simple fronds, less than 1 mm wide. Hairs present within the sporangia.

Schizaea **spp.**
1. **Ecol.**: Terrestrial in lower montane ridge forest at 800-850 m altitude.

Specimens: ARFAK MTS.: Mokwam Trail - *Johns* 7712.
Notes: Stems to 30 cm long.

2. **Ecol.**: Terrestrial in partial shade in disturbed lowland rain forest at 0-20 m altitude.
Specimens: ARFAK PLAINS: Near road to Sungai Wariori - *Sands* 6259.
Notes: Plants to 25 cm tall, stipes brown.

3. **Ecol.**: Terrestrial in partial shade on dry river bank in lower montane forest at 625 m altitude.
Specimens: ARFAK PLAINS: On steep ridges to G. Itsiwei - *Sands* 6355.
Notes: Plants to 60 cm tall, stipes brown.

4. **Ecol.**: Terrestrial along ridge in lower montane forest on limestone escarpment at 900 m altitude.
Specimens: KEBAR VALLEY: E of Andjai 2 - *Johns* 8277.
Notes: Stems to 35 cm long. Fronds to 11 cm long, sori apical in 8's.

5. **Ecol.**: Scarce terrestrial in forest at 5-20 m altitude.
Specimens: SIDAI: W of Manokwari - *Chr. Koster* BW 6827.

THELYPTERIDACEAE Ching

Thelypteridaceae Pichi Sermolli
Sphaerostephanaceae Ching nom. inval.

Copeland, E.B. (1949). Aspidiaceae of New Guinea. *Philip. J. Sci.* **78**: 389-475.

Holttum, R.E. (1969). Studies in the Theypteridaceae. The genera *Phegopteris, Pseudophegopteris* and *Macrothelypteris*. *Blumea* **17**: 5-32.

Holttum, R.E. (1971). Studies in the family Thelypteridaceae III. A new system of genera in the Old World. *Blumea* **19**: 17-52.

Holttum, R.E. (1972). Studies in the family Thelypteridaceae IV. The genus *Pronephrium* Presl. *Blumea* **20**: 105-126.

Holttum, R.E. (1973). Studies in the Thelypteridaceae V. The genus *Pneumatopteris* Nakai. *Blumea* **21**: 293-325.

Holttum, R.E. (1974). Studies in the family Thelypteridaceae VII. The genus *Chingia*. *Kalikasan* **3**: 13-28.

Holttum, R.E. (1975). Studies in the Thelypteridaceae VIII. The genera *Mesophlebion* and *Plesioneuron*. *Blumea* **22**: 223-250.

Holttum, R.E. (1976a). Studies in the Thelypteridaceae V. The genus *Coryphopteris*. *Blumea* **23**: 18-47.

Holttum, R.E. (1976b). The genus *Christella* Leveille Sect. *Christella*. Studies in the family Thelypteridaceae XI. *Kew Bull.* **31**(2): 293-339.

Holttum, R.E. (1977a). Studies in the Thelypteridaceae XII. The genus *Amphineuron* Holttum. *Blumea* **22**: 205 - 218.

Holttum, R.E. (1977b). The family Thelypteridaceae in the Pacific and Australasia. *Allertonia* **1**: 169 - 243. f. 1-10.

Holttum, R.E. (1981). Thelypteridaceae. *Flora Malesiana Ser. II.* **1**: 331 - 560, f. 1 - 20.

Holttum, R.E., Sen, U. & Mittra, D. (1970). Studies in the family Thelypteridaceae II. A comparative study of the type species of *Thelypteris* Schmidel, *Cyclosorus* Link and *Ampelopteris* Kunze. *Blumea* **18**: 195 - 215.

Thelypteridaceae spp.

1. **Ecol.**: In disturbed lowland rain forest.
Specimens: *Johns* 8179.
Notes: Plants to 1.5-2 m tall. Rhizome yellowish, densely covered with yellow-brown scales.

2. **Ecol.**: In ridge forest at 800-850 m altitude.
Specimens: ARFAK MTS.: *Johns* 7684.
Notes: Trunk to 15 cm long, covered with large, scattered, brown scales. Fronds pinnate-pinnatifid. Veins free.

3. **Ecol.**: In ridge forest along track from Wamare to Mokwam at 800-850 m altitude.
Specimens: ARFAK MTS.: Mokwam Trail - *Johns* 7686.
Notes: Caudex with broad, brown, long-pointed scales. Stipe to 100 cm long.

4. **Ecol.**: In montane forest on uplifted coral at 300 m altitude.
Specimens: *Johns* 8117.
Notes: Plants to 2 m tall. Rhizome forming a short pseudo-trunk. Stipe to 70 cm long.

5. **Ecol.**: Terrestrial in disturbd lowland tropical rain forest at 50 m.
Specimens: *Johns* 8152.
Notes: Rhizome long-creeping, densely covered with black hairs. Fronds 1-2 m tall.

6. **Ecol.**: Terresrial on slope near track at 1950 m altitude.
Specimens: *Johns* 8096.
Notes: Rhizome short-creeping. Stipe to 8.5 cm, densely scaly. Rachis to 26 cm long. Pinnate-pinnatifid.

7. **Ecol.**: Terrestrial in primary forest at 70 m altitude.
Specimens: ARFAK PLAINS: Wariori Flood Plain - *Johns* 8211.
Notes: Caudex to 30 cm, apex with dense brown scales. Fronds pinnate, stipe to 1 m. See table.

8. **Ecol.**: Terrestrial in primary forest at 70 m altitude.
Specimens: ARFAK PLAINS: Wariori Flood Plain - *Johns* 8214.
Notes: Rhizome short-creeping. Fronds pinnate-pinnatifid, Stipe to 41 cm with 7 small leaflets to 1 cm long.

9. **Ecol.**: Terrestrial in ridge top forest on limestone escarpment at 900 m altitude.
Specimens: KEBAR VALLEY: On ridge above Apni Valley east of Andjai 2 - *Johns* 8257.
Notes: Rhizome long-creeping, black. Stipe to 1.2 m long, grooved in several rows.

10. **Ecol.**: Terrestrial in *Araucaria* plantation at 600 m altitude.
Specimens: *Johns* 8330.
Notes: Rhizome long-creeping with many long scales. Stipe purplish-black, to 96 cm long, with 8 very reduced pinnae.

11. **Ecol.**: Terrestrial in disturbed hill forest on limestone outcrop at 600 m altitude.
Specimens: *Johns* 8359.
Notes: Stipes to 70 cm long with scattered small pinnae in upper sections. Fronds pinnate.

12. **Ecol.**: Growing in very wet swamp area, rooted in water. From 1700-1940 m altitude.
Specimens: *Johns* 7949.
Notes: Rhizome short-creeping in water.

AMPHINEURON Holttum

Amphineuron immersum (Blume) Holttum in Nayar & Kaur, Comp. to Beddome 203 (1974).
Aspidium immersum Blume *Enum. Pl. Jav.* 156 (1828).
Dist.: Assam/Hainan to Malesia (throughout) to Queensland, Vanuatu and New Caledonia.
Ecol.: Gunung Meja: In secondary forest on uplifted coral along road from Manokwari at 110 m altitude.
Specimens: *Johns* 7662.
Notes: Trunk to 6-8 cm. Stipe to 85 cm with double groove abaxially. Pinnae to 27 on each side. Veins free. Sori marginal indusiate.

CHINGIA Holttum

Chingia imponens (Ces.) Holttum *Kalikasan* **3**: 21 (1974). Holttum (1981: 391).
Polypodium imponens Cesati **Type**: *Beccari* 12668. Arfak Mts. (FI).
Dryopteris imponens (Cesati) C.Chr.
Phegopteris imponens Alderw. (1909).
Dist.: Holttum states endemic to E NG.
Specimens: **ARFAK MTS.**: *Beccari* 12668. (FI).

CHRISTELLA Leveille

Christella **sp.**
Specimens: *van Royen & Sleumer* 6730 (L).

CORYPHOPTERIS Holttum

Coryphopteris badia (Alderw.) Holttum *Blumea* **23**: 44 (1976). Holttum (1981: 369).
Dryopteris badia Alderw. *Bull. Jard. Bot. Btzg.* II. **16**: 9 (1914). **Type**: Mathew 674. Sumatra. Mt Tandikat. (BO: E).
Dryopteris villosipes Gepp in Gibbs 70 (1917). **Type**: L.S. Gibbs 5627. Arfak Mts. (BM).
Dist.: Sumatra to New Guinea.
Ecol.: In mossy forest from 1400-2500 m altitude.
Specimens: KOEBRE MT.: *Gibbs* 5627 (**Type**).

Coryphopteris **sp. nov. 1** *aff.* ***badia*** (Alderw.) Holttum
Ecol.: In subalpine forest at 2100 m altitude.
Specimens: KOEBRE MT.: *Johns* 8062.

***Coryphopteris* sp. nov. 2**
Ecol.: Arfak Mts at 2350 m.
Specimens: ARFAK MTS.: *Craven & de Fretas* 27 (K).

***Coryphopteris* sp. 3**
Specimens: *van Royen & Sleumer* 8223 B (?).
Notes: See 8223 in *Thelypteris*.

***Coryphopteris* spp.**
Specimens: *Croxall* 7; *Croxall* 27; Croxall 1B; M.A. Hildebraed 1. N Arfaks 2480 m. (K).

CYCLOSORUS Link.

***Cyclosorus* sp. 1**
Ecol.: 1860 m.
Specimens: *E.A. Widjaja* 4382.

***Cyclosorus* sp. 2**
Ecol.: 30 m.
Specimens: *E.A. Widjaja* 4625.

MACROTHELYPTERIS (H.Ito) Ching

Macrothelypteris polypodioides (Hook.) Holttum, *Blumea* **17**: 29 (1969). Holttum (1977: 179).
Alsophila polypodioides Hook. in Nightingale *Oceanic Sketches* 131 (1835). **Type**: Nightingale. 'South Sea Islands'.
Cheilanthes gigantea Cesati *Rendic. Ac. Sci. Nap.* **16**: 25, 29 (1877). **Type**: Beccari. Mt Arfak. Irian Jaya. (FI: K).
Dist.: Thailand throughout Malesia to W Pacific.
Ecol.: In open areas at medium altitudes.
Specimens: BUKIT BORAI: *Sands* 6706. RANSIKI RD.: *Johns* 7832.

Macrothelypteris torresiana (Gaud.) Ching
Dist.: Mascarene Islands to mainland Asia, throughout Malesia to Queensland, Hawaii and Polynesia.
Ecol.: Usually at low altitudes. In wet ditch in *Araucaria* plantation.
Specimens: KEBAR VALLEY: Between Kampong Anjai and K. Akmuri. 600 m altitude - Johns 8336.

MENISCIUM Schreb.

***Meniscium* sp. 1**
Specimens: **ANDAI**: *Beccari* s.n.

PARATHELYPTERIS (H.Ito) Ching

Parathelypteris beddomei (Baker) Ching *Acta Phytotax. Sinica* **8**: 302 (1963).

Nephrodium beddomei Baker *Syn*. 267 (1867). **Type**: Beccari s.n. Andai. Arfak Mts. (FI: K).

Lastrea beddomei (O.Kuntze) Gibbs (1917: 70).

Dryopteris beddomei (Baker) O.Ktze. *Rev. Gen. Pl.* **2**: 812 (1891).

Dist.: India and Ceylon, S China to Philippines and New Guinea.

Ecol.: Common in open marsh near lake.

Specimens: ANGGI LAKES: *Gibbs* 5939 (BM).

PNEUMATOPTERIS Nakai

Pneumatopteris papuana Holttum *Blumea* **21**: 311 (1973). **Type**: L.J. Brass 22778. Mt Dayman, E New Guinea. 2000 m. (LAE: BM, L).

Ecol.: Terrestrial in *Araucaria* plantation at 600 m altitude.

Specimens: *Johns* 8337.

Notes: Rhizome forming a small trunk with abundant small black scales. Bases of stipes, and croziers, with similar scales.

First record from Irian Jaya.

Pneumatopteris* sp.** matching ***Pneumatopteris angustisquamata Holttum ined. Carr 13632 Papua. (K).

Specimens: ARFAK MTS.: Mupi Dessa - trail from Mupi Village to G. Humibou - *Sands* 6813.

PRONEPHRIUM C.Presl

Pronephrium beccarianum (Ces.) Holttum *Blumea* **20**: 107 (1972). Holttum (1981: 513).

Meniscium beccarianum Cesati, *Rendic. Acad. Napoli* **16**: 27, 30 (1877). **Type**: Beccari. Irian Jaya, Andai. (Herb. Becc. 12727).

Dist.: Moluccas to New Guinea, Solomon Islands and Fiji.

Ecol.: Terrestrial in ridgetop forest forest dominated by *Araucaria*, *Lithocarpus*, *Litsea* etc. at 550-650 m altitude.

Specimens: ARFAK MTS.: Mokwam Trail - *Johns* 7774. ARFAK PLAINS: On track to G. Itsiwei - *Sands* 6433.

Notes: Terrestrial with stipes to 5-7 cm long. Fronds simple to 25x4.5 cm. Young fronds reddish. Frond margin transulcent. Plants to 45 cm tall. Stipes and costa dark maroon to chocolate-brown.

***Pronephrium* sp.**

1. **Ecol.**: Terrestrial in valley floor forest at 600-650 m altitude.
   Specimens: **ARFAK MTS.**: Mokwam Trail - *Johns* 7767.
   Notes: Terrestrial with stipes to 10-12 cm long.

2. **Specimens**: *Croxall* 25.

SPHAEROSTEPHANOS J.Smith

Sphaerostephanos arfakianus (Baker) Holttum *Fl. Malesiana* ser. II. **1**(3): 484 (1981).

Polypodium arfakianum Baker in Beccari *Malesia* **3**: 45 (1880). **Type**: Beccari s.n. 1872. Putat, Mt Arfak. Irian Jaya. (FI: K).

Dryopteris sepikensis Brause *Bot. Jahrb.* **56**: 101 (1920). Copeland (1951: 443). **Type**: Ledermann 12053. Schraderberg. NE New Guinea. (B).

Dist.: Moluccas to New Guinea. Widely dispersed.

Ecol.: Terrestrial on ridge in lower montane forest dominated by *Araucaria-Lithocarpus-Litsea* forest at 650(-2000) m altitude. Also in disturbed lowland tropical rain forest at 50 m.

Specimens: *Johns* 7776; *Johns* 8160; *Beccari* 12659. *Johns* 8195. *Sands* 6443.

Notes: Rhizome with flattened brown scales. Stipes to 9 cm. Rachis to 140 cm with c. 36 pairs of pinnae.

Caudex to 10-15 cm tall. Fronds pinnate, stipe black, to 20 cm, bases with large brown scales.

An exindusiate form ??

Sphaerostephanos polycarpus (Bl.) Copel. *Un. Calif. Publ. Bot.* **16**: 60 (1929). Copeland (1960: 379). Holttum (1981: 448).

Aspidium polycarpum Blume *Enum. Pl. Jav.* 156 (1828). **Type**: Blume. Noesa Kambangan, Java. (L).

Dist.: Thailand, throughout Malesia to Solomon Islands and Samoa.

Ecol.: On forest floor in shade. Partially disturbed lowland rain forest. In open 0-20(-1500) m.

Specimens: ARFAK PLAINS: Wariori Flood Plain - *Sands* 6180.

Notes: Fronds to 3 m tall. Stipe reddish-brown at base with brown scales.

Sphaerostephanos unitus (L.) Holttum *J. S. Af. Bot.* **40**: 165 (1974). Holttum (1975: 63).

Polypodium unitum L. *Syst. Nat.* ed. 2 **10**: 1326 (1759).

Dryopteris unita (L.) Kuntze

Dist.: E Africa to India through to Malesia. Also Queensland.

Ecol.: Usually at low altitudes in damp areas. Terrestrial in partial shade on stony ground. On old lake bed.

Specimens: KEBAR VALLEY: Bukit Sontiri - *Sands* 6560. *van Royen & Sleumer* 6783.

Notes: Fronds erect, arising from long-creeping rhizome.

Sphaerostephanos unitus (L.) Holttum var. ***mucronatus*** (Christ) Holttum *Fl. Malesiana* ser. II. **1**(3): 478 (1981).

Dryopteris cucullata var. *mucronata* Christ *Philip. J. Sci.* **2**: 195 (1907). Alderw. (1908: 819). **Type**: Cuming 182. Luzon. (FI, K, LE).

Dist.: Borneo and Philippines, Sulawesi, Moluccas, New Guinea to Polynesia.

Ecol.: Terrestrial along Ransiki Road in lowland rain forest on alluvial plain at 0-100 m altitude.

Specimens: RANSIKI ROAD: *Johns* 7843; *Johns* 7905.

Notes: Stipes, black, heavily grooved, to 85-100 cm tall, reddish, with scattered very small pinnae at wide intervals.. Base of stipes without scales. Rachis to 950 cm. Basal pinnae falling leaving base on rachis. Fronds pinnate, pinnae to 17x2 cm.

Sphaerostephanos unitus (L.) Holttum var. ***papilliferus*** Holttum *Fl. Malesiana* ser. II. **1**(3): 479 (1981). **Type**: Hoogland 9084. Huon Peninsula. N.E. New Guinea. 1320 m. (K).

Dist.: New Guinea. Many collections.

Ecol.: Terrestrial in disturbed areas around airstrip. From (1300-)1700-1940(-2000) m altitude.

Specimens: IRAY: *Johns* 7906. ANGGI LAKES: *Johns* 7971.

Notes: Terrestrial with long-creeping rhizomes on surface of soil. Stipes to 80 cm long, rachis to 64 cm. Pinnae to 15 x 1.5 cm. Indusium hairy.

***Sphaerostephanos* sp. nov. 1**

Dist.: Known only from a single collection

Ecol.: On steep ridge in hillforest. Terrestrial in shade. 500 m. alt.
Specimens: ARFAK MTS.: Mokwam Trail - *Sands* 6135.
Notes: This species has obvious sessile glands between the veins on the adaxial surface. Sporangia are glabrous. It is superficially similar to *S. pilosquamatus* Holttum (sp. 82 from Bismark Arch.). but the latter species has no glands adaxially on lamina and differs in that many of the sporangia have setae. The surface of this species is pustular, as in both *Pneumatopteris* and *Pronephrium*, but Holttum (1982) noted that some species of *Sphaerostephanos* are pustular.

THELYPTERIS L.

***Thelypteris* spp.**
Specimens: *van Royen & Sleumer* 6871; *van Royen & Sleumer* 7012; *van Royen & Sleumer* 8223 A; *van Royen & Sleumer* 8226; *van Royen & Sleumer* 5090.

VITTARIACEAE

ANTROPHYUM Kaulf.

***Antrophyum* spp.**

1. **Ecol.**: Epiphyte in mixed lower montane rain forest on ridge dominated by *Araucaria*, *Lithocarpus* and *Litsea*.
Specimens: ARFAK MTS.: Mokwam Trail - *Johns* 7784.
Notes: Rhizome short-creeping. Sori forming long line, not continuous.

2. **Ecol.**: Rare trunk epiphyte in swampy bamboo community at 650 m altitude.
Specimens: KEBAR VALLEY: S of Andjai 2 - *Johns* 8288.
Notes: Fronds to 40 x 10 cm. Lacking stipe. Sori copiously anastomising.

3. **Ecol.**: Epiphyte on tree trunks in ridge top forest along limestone escarpment at 900 m altitude.
Specimens: KEBAR VALLEY: E of Andjai 2 - *Johns* 8272.
Notes: Fronds simple, to 22 cm long. Sori longitudinally along veins, not or rarely anastomising.

4. **Ecol.**: Epiphyte on tree trunk in mixed lowland rain forest on flood plain near Wariori River at 70 m altitude.
Specimens: ARFAK PLAIN: Wariori River Flood Plain - *Johns* 8237.
Notes: Rhizome short-creeping. Fronds to 32 x 4 cm. Sori forming short line to 1 cm long, not continuous.

5. **Ecol.**: Epiphyte on tree boles 2-3 m above ground in *Araucaria* plantations at 600 m altitude.
Specimens: KEBAR VALLEY: E of Andjai 2 - *Johns* 8272.
Notes: Rhizome short-creeping. Fronds to 33 x 10 cm. Stipe very short to lacking. Sori immersed in the thick fleshy fronds. Sori free on smaller fronds.

6. **Ecol.**: Small epiphyte in forest near hot springs at 600 m altitude.
Specimens: KEBAR VALLEY: S of Andjai 2 - *Johns* 8370.

Notes: Rhizome covered with dense brown scales. Fronds to 23 x 4 cm, stipes short.

VITTARIA J.E.Smith

Vittaria crassifolia Baker. Gepp in Gibbs (1917: 74);
Dist.: Borneo to New Guinea.
Ecol.: Epiphytic on forests. 2400 m.
Specimens: KOEBRE MT. : *Gibbs* 5729 (BM);

Vittaria elongata Sw. Gepp in Gibbs (1917: 74);
Dist.: Tropical Asia, to E Australia and Polynesia.
Ecol.: Epiphytic on forests

***Vittaria* spp.**

1. **Ecol.**: Pendant epiphyte in valley floor forest along trail from Wamare to Mokwam at 600-650 m altitude.
Specimens: ARFAK MTS.: Mokwam Trail - *Johns* 7753.
Notes: Rhizome short-creeping. Fronds to 1.3 m long.

2. **Ecol.**: Small epiphyte on trunk of *Pandanus* in ridge top forest along limestone escarpment at 900 m altitude.
Specimens: KEBAR VALLEY: E of Anjai 2, Apni Valley - *Johns* 8261.
Notes: Fronds simple to 20-25 cm long.

3. **Ecol.**: Epiphyte on climber in lowland rain forest at 6-7 m height above ground, at 70 m altitude.
Specimens: ARFAK PLAINS: Wariori Flood Plains - *Johns* 8218.
Notes: Stipe short, to 3 cm long. Fronds simple, to 54 x 3.5 cm. Sorus some 2 mm inside margin.

4. **Ecol.**: Epiphyte on trunk in subalpine forest at 1950 m altitude.
Specimens: KOEBRE MT.: Track from Iray to summit - *Johns* 8088.

5. **Ecol.**: Pendent epiphyte in ridge top, lower montane forest with *Araucaria*, *Lithocarpus* and *Litsea* 650 m altitude.
Specimens: ARFAK MTS.: Mokwam Trail - *Johns* 7772.
Notes: Fronds to 114 cm long. 0.7 - 0.8 cm across.

6. **Ecol.**: Pendent epiphyte in ridge top, lower montane forest with *Araucaria*, *Lithocarpus* and *Litsea* 650 m altitude.
Specimens: ARFAK MTS.: Mokwam Trail - *Johns* 7773.
Notes: Fronds to 70 cm long. 2.2 cm across.

7. **Ecol.**: Pendent epiphyte in ridge top, lower montane forest with *Araucaria*, *Lithocarpus* and *Litsea* 650 m altitude.
Specimens: ARFAK MTS.: Mokwam Trail - *Johns* 7806.
Notes: Fronds to 33 x 4 cm. Stipes short.

8. **Ecol.**: Small epiphyte in forest near hot springs at 600 m altitude.

Specimens: KEBAR VALLEY: S of Anjai 2 - *Johns* 8368.
Notes: Fronds 26-27 x 0.6-0.7 cm. Stipes short.

WOODSIACEAE (Dryopteridaceae)

Copeland, E.B. (1949). Aspidiaceae of New Guinea. *Philip. J. Sci.* **78**: 389-475.

DIPLAZIUM Swartz

Diplazium cordifolium Bl.
Ecol.: In partial shade in ridge top lower montane forest. 635 m alt.
Specimens: ARFAK PLAINS: Ridge to G. Itsiwei - *Sands* 6364.
Notes: Plants to 75 cm tall, stipes 25-40 cm. Fronds leathery.

***Diplazium* spp.**

1. **Ecol.**: In deep shade in lowland rain forest on very damp ground. 30-50 m alt.
   Specimens: ARFAK PLAINS: Foothills along S margin - *Sands* 6320.
   Notes: Erect fern with fronds to 35 cm long, 2-3 pinnae of each side plus terminal pinna.

2. **Ecol.**: In partial shade in ridge top lower montane vegetation with gaps. 635 m.
   Specimens: ARFAK PLAINS: On steep ridges to G. Itsiwei - *Sands* 6368.
   Notes: Fronds erect to 1 m tall with stipes 75 cm. 6 pinnae.

3. **Ecol.**: Track from Warmare to Mokwam in ridge forest at 800-850 m altitude.
   Specimens: ARFAK MTS.: Mokwam Trail - *Johns* 7716.
   Notes: Fronds pinnate-pinnatifid with a deep groove, confluent with groove in secondary rachis. Veins free.

4. **Ecol.**: Track from Warmare to Mokwam in valley floor forest at 600-650 m altitude.
   Specimens: ARFAK MTS.: Mokwam Trail - *Johns* 7765.
   Notes: Stipes to 1 m long, spiny on all surfaces. Fronds pinnate-pinnatifid, rachis to 1.5-2 m long.

***Diplazium* spp.**
Schram BW 7991 (L); *Main* 381 (L); *Djamhari* 531 (L); *Johns* 8122; *Johns* 8173; *Johns* 8203; *Johns* 8202.

Alsophila angiensis Gepp. in Gibbs, *Phytogeog. and Flora Arfak Mts.* 68 (1917). **Type**: L.S. Gibbs 5968. 12/1914. Anggi Lakes. Tree fern to 3 m tall. 2550m. (Holotype: BM)

Alsophila arfakensis Gepp. in Gibbs, *Phytogeog. and Flora Arfak Mts.* 68 (1917). **Type**: L.S. Gibbs 5990. Dec. 1914. Anggi Lakes. Arfak Mts. Common in undergrowth of moss forest. 2100 - 2550 m. (Holotype: BM).

Asplenium linza Cesati *Rendic. Accad. Sc. Fil. Mat. Napoli* **16**: 26, 29 (1877). **Type**: Beccari s.n. Ramoi. Arfak Mountains. (FI). ***Triphlebia linza*** (Cesati) Baker in Beccari *Malesia* **3**: 42 (1886). = ***Diplora d'urbillaei***

Cyathea arfakensis Gepp in Gibbs, *Phytogeog. and Flora Arfak Mts.* 68 (1917). **Type**: L.S. Gibbs 6008. 12/1914. Anggi Lakes, on SW ridge. Undergrowth in forest near lake. 2400m. (Holotype: BM).

Davallia longipinnula Cesati *Rendic. R. Accad. Sci. Fis. Mat. Napoli* **16**: 29 (1877). **Type**: *Beccari s.n.* Ramoi New Guinea. (Holotype FL). ***Tapeinidium longipinnulum*** (Cesati) C.Chr. *Ind. Fil. Suppl.* **3**: 176 (1934). Kramer (1968: 555); Kramer (1971: 197).

Hymenophyllum cernuum Gepp. in Gibbs, *Phytogeog. and Flora Arfak Mts.* 68 (1917). **Type**: *L.S. Gibbs* 5964. 12/1914. Anggi Lakes, near Female Lake. Epiphyte in forest patch. 2100m. (Holotype: BM)

Hymenophyllum cincinnatum Gepp. in Gibbs, *Phytogeog. and Flora Arfak Mts.* 68 (1917). **Type**: *L.S. Gibbs* 5989. 12/1914. Anggi Lakes. Epiphyte in moss-forest. 2550m. (Holotype: BM)

Lastrea villosipes Gepp. in Gibbs, *Phytogeog. and Flora Arfak Mts.* 70 (1917). **Type**: *L.S. Gibbs* 5627. 12/1914. Koebre Mt. Epiphyte in moss-forest. 2100-2400m. (Holotype: BM)

Gymnogramme pteridiformis Cesati *Rend. Acad. Napoli* **16**: 30 (1877). **Type**: *Beccari s.n.* Andai. (Holotype Fl. Herb. Beccari 12704: K). ***Thysanosoria pteridiformis*** (Cesati) C.Chr. *Ind. Fil. Suppl.* **3**: 187 (1934).

Polybotrya arfakensis Gepp. in Gibbs Arfak (1917: 71). **Type**: L.S. Gibbs 5984. Anggi Lakes. 2100 m. (BM). ***Cyathea biformis*** (Rosenst.) Copel.

Polypodium (*Dictyopteris*) ***andaiense*** Baker in Beccari, *Malesia* **3**: 45 (1886). **Type**: Beccari s.n. Andai, Irian Jaya. (FI: K). ***Pleocnemia andainensis*** (Baker) Holttum *Kew Bull.* **29**: 350 (1974).

Thysanosoria dimorphophylla Gepp in Gibbs *Dutch NW New Guinea* 193, pl. 4 (1917). **Type**: *L.S. Gibbs* 6162. (Holotype BM). ***Thysanosoria pteridiformis*** (Cesati) C.Chr. *Ind. Fil. Suppl.* **3**: 187 (1934).

Triphlebia dimorphophylla Baker in Beccari *Malesia* **3**: 42 (1886). Brause (1920: 144). **Type**: Beccari s.n. Andai, Arfak Mountains. (FI). = ***Diplora d'urbillaei***

GYMNOSPERMS

R.J. Johns

CYCADACEAE

CYCAS L.

Hill, R.D. (1992). A preliminary account of *Cycas* (Cycadaceae) in Queensland. *Telopea* 5(1): 177-206.

Hill, R.D. (1995a). The *Cycas media* group (Cycadaceae) in New Guinea. *Australian Systematic Botany* 7: 527-541.

Hill, R.D. (1995b). The genus *Cycas* (Cycadaceae) in in the Indian region, with notes on the application and typification of the name *Cycas circinalis* L. *Taxon* **44**: 23-31.

Cycas rumphii Miq., *Bull. Sci. Phys. Nat. Neerl.* **2**: 45 (1839). Hill (1992: 183); Hill (1995a: 549). Hill (1995b: 27). **Type**: Rumphius Herb. Amboin. **1**. 86, t. 22, 23 (1749).

Ecol. In grassland areas at low altitudes.

Dist.: Widespread in SE Asia and Pacific.

Specimens: KAPAOR: *Kanehira & Hatusima* 13152. WAREN: *Beccari* 58; *Kanehira & Hatusima* 13151; *Kanehira & Hatusima* 11983.

Notes:

Cycas circinalis L., as now understood, is an Indian endemic (Hill 1995: 29). The name should not be applied to any specimens from Irian Jaya.

Cycas apoda K.D. Hill is reported from the Vogelkop but no specimens are cited from Manokwari Kebupatan (Hill 1995a: 553).

CUPRESSACEAE

A single genus and species in the Kepala Burung. The species represented by two varieties.

PAPUACEDRUS H.J.Li

Johns, R.J. (1995). 266. *Papuacedrus papuana* var. *papuana*. *Curtis's Botanical Magazine* **12**(2): 66 - 72.

Laubenfels, D.J. de (1988). Cupressaceae (in Coniferales). *Flora Malesiana* Ser. 1. Vol. **10**: 442-447.

Papuacedrus papuanus (F.Muell.) H.J.Li, *J. Arnold Arbor.* **34**: 25, f.33 (1953). **Type**: MacGregor s.n. Mt Knutsford, Owen Stanley Range. Papua New Guinea.

Libocedrus papuana F.Muell. *Trans. Roy. Soc. Victoria* **1**(1): 32 (1889).

Thuja papuana (F.Muell.) Voss.

Notes: Intermediate foliage occurs between the adult plants of the two varieties but the juvenile foliage is quite distinct

Papuacedrus papuanus (F.Muell.) H.J.Li **var.** ***papuanus***

Ecol.: At low altitudes with Araucaria and Dacrydium (620 m) in the Cycloop Mountains and in the Kebar Valley at 870 m. Common in forest above 1560 m, in fagaceous forest, mid montane to subalpine conifer forest. As a scattered emergent at higher altitudes in sub-alpine shrubery - Nettoti Range above 3000 m altitude.

Dist.: Throughout New Guinea

Specimens: NETTOTI MT.: *Davis* 824. ARFAK MTS.: *Kostermans* 2172; *Kostermans* 2271; *Kostermans* 2308. MT. KOEBRE: *Johns* 8084. WITHOUT LOCALITY: BW 3281; BW 10410; BW 12863; BW 12867. *van Royen & Sleumer* 8227.

Papuacedrus papuanus (F.Muell.) H.J.Li, **var. *arfakensis*** (Gibbs) R.J. Johns *Curtis's Botanical Magazine* **12**(2): 70. **Type**: Gibbs 5594. Koebre Ridge, Arfak Mts.

Libocedrus papuanus var. *arfakensis* (Gibbs) de Laub.

Libocedrus arfakensis Gibbs *Contr. Phytogeog. Fl. Arfak Mts.* 84, f. 6a-6b (1917).

Papuacedrus arfakensis (Gibbs) H.J. Li

Ecol.: Rare to locally common on ridges in Kebar Valley at 900-1130 m, and in heath vegetation on Mt Nettoti at 1980 m. Also common around Anggi Lakes at 1850-2300 m.

Dist.: Common throughout Irian Jaya, extending to Moluccas.

Specimens: ARFAK MTS.: *Gibbs* 5500; *Gibbs* 5550; *Kanehira & Hatusima* 13467; *Kanehira & Hatusima* 13716; *Kanehira & Hatusima* 14035. ANGGI LAKES: BW 254; BW 260; BW 279; BW 2009. KOEBRE MT.: *Gibbs* 5594. WITHOUT LOCALITY: BW 5595; BW 6374; BW 6866; BW 7973; *van Royen & Sleumer* 3874.

ARAUCARIACEAE

Laubenfels, D.J. de (1988). Araucariaceae (in Coniferales). *Flora Malesiana* Ser. 1. Vol. **10**: 419-442.

ARAUCARIA Juss.

Araucaria hunsteinii has been listed in some books from Irian Jaya, probably due to misidentification in the field. No herbarium material is known from Irian Jaya. *A. cunninghamii* is grown in plantations in Irian Jaya.

Araucaria cunninghamii Ait. ex D.Don in Lamb. **var. *papuana*** Laut., *Bot. Jahrb.* **50**: 51 (1913). de Laub. (1988: 425).

Araucaria papuana

Araucaria beccari Warb.

Ecol.: Common on ridges in Lower Montane fagaceous forest.

Dist.: Widespread in New Guinea. Scattered in isolated patches, locally common.

Specimens: KEBAR VALLEY: Not collected. ARFAK MTS: *Beccari* 9804; *Beccari* 9805; *Beccari* 9806; *Gibbs* 5934. ANGGI LAKES: BW 287; BW 2047; *Gibbs* 5748; Gibbs *5749*. WAREN: *Kanehira & Hatusima* 13178. WITHOUT LOCALITY: BW 710; BW 2313; BW 2316; BW 2317; BW 2341; BW 2348; BW 5749; BW 7945; BW 12915; *van Royen & Sleumer* 5016.

Notes: The type variety, **var. *cunninghamii***, is restricted to Eastern Australia.

Araucaria hunsteinii K.Sch., Fl. Kaiser Wilhelms Land 11, t.4, f.8 (1889). de Laub. (1988: 424).

A. klinkii is a variety of this species which is very glaucous.

Ecol. Emergent in lower montane oak forests. Sometimes forming pure stands.

Dist. Endemic to Papua New Guinea: not reliably reported from Irian Jaya.

Specimens: Cultivated. Not collected.

Notes: MANOKWARI: Gunung Meja - *Widjaja & Hamzah* 3122 which is identified as ***A. bidwilii*** is possibly ex. cultivation.

AGATHIS Salisb.

This genus can be confused with *Nageia* (Podocarpaceae). See notes there to distinguish the genera. *Agathis* leaves can also be distinguihed by the presence of a resinous exudate when broken.

Agathis labillardieri Warb., *Monsunia* **1**: 183 (1900). de Laub. (1988: 441).
Ecol.: From sea-level to 1300 m, often on ultrbasic soils. Very common to locally dominant in Irian Jaya.
Dist.: Occurs in the western parts of New Guinea, east of the Sepik Valley foothills.
Specimens: WITHOUT LOCALITY: BW 2315; BW 3283; BW 7963; BW 12823; *van Royen & Sleumer* 7116.

PHYLLOCLADACEAE

PHYLLOCLADUS L.C. Rich *ex* Mirbel

Laubenfels, D.J. de (1988). *Phyllocladus* (in Coniferae). *Flora Malesiana* Ser. 1. Vol. **10**: 355-360.

This genus is easily recognised by its thick phyllodes which replace the leaves.

Phyllocladus hypophyllus Hook. f. *Icon. Pl.* n.s. **5**: t. 889 (1852). de Laub. (1988: 355).
Phyllocladus major Pilger *Bot. Jahrb.* **54**: 211 (1916).
Phyllocladus protractus (Warb.) Pilger
Ecol.: In moist montane forests from 900 to 3200 (-4000) m. Scattered at lower altitudes, but can be dominant to co-dominant in forests at higher altitudes.
Dist.: Philippines, Boreno, Sulawesi, Moluccas to New Guinea. Widespread along the central mountain ranges.
Specimens: MT NETTOTI: *van Royen* 3873; *van Royen & Sleumer* 7403; *Versteegh* BW 10407. KEBAR VALLEY: Neentjapaki - *Kalkman* BW 6373; Andjar - *Koster* BW 6887; Tobie Mts. - *Schram* BW 7972. ANGGI LAKES: *Gibbs* 5992; *Versteegh* BW 248, BW 253, BW 281; *Kanehira & Hatusima* 13704, *Kanehira & Hatusima* 14906; *Stefels* BW 2008, BW 2010, BW 2031, BW 2254. Lake Gita - *Kanehira & Hatusima* 13460. KOEBRE RIDGE: *Gibbs* 5657. ARFAK MTS.: *Kostermans* 2270. RANSIKI: *Mangold* BW 2262; Mt Mundi - *Mangold* BW 2254. WITHOUT LOCALITY: *Gibbs* 6127; BW 7972; BW 10326; *van Royen* 3873.

PODOCARPACEAE

Johns, R.J. (1982). Common Forest Trees of Papua New Guinea. Pt. 1. Gymnosperms. P.N.G. Univ. of Technology. Lae.
Laubenfels, D.J. de (1969). A revision of the Asian and Pacific Rainforest Conifers. *J. Arn. Arbor.* **50**(2-3): 274-369.
Laubenfels, D.J. de (1988). Coniferae. *Flora Malesiana* Ser. 1. Vol. **10**: 442-447.

The following record (n.v.) was published for the family from the Kepala Burung.
Podocarpaceae sp. *Beccari* 946 (FL).

KEY TO GENERA OF PODOCARPACEAE IN KEPALA BURUNG

(Based on Johns 1982)

1. Leaves with many, often indistinct, parallel veins.. *Nageia*
1. Leaves with a single distinct midrib...2

 2. Seeds covered by a scale or fused to it...3
 2. Seeds free, projecting above the fertile scale...4

3. Common canopy tree in wet areas. Leaves needle-like, to 2 cm long. Bark shed in irregular flakes.. *Dacrydium*
3. Common understorey tree. Leaves stiff, prickly. Bark irregularly fissured.... *Falcatifolium*

 4. Leaves flat, scale-like (to awl-shaped - then closely appressed to branch). Common in higher altitude forest (above 1500 - 2000 m)................ *Dacrycarpus*
 4. Leaves to 4-5 cm long, or longer. Common in forest at low and medium altitudes..5

5. Bark peeling in long , narrow strips. Seed single, seated on a red, fleshy, receptacle, to c. 1cm diam.. *Podocarpus*
5. Bark peeling in large, irregular flakes. Seeds naked, orange-red, to 2.5 cm diam. .... *Prumnopitys*

DACRYCARPUS (Endl.) de Laub.

Dacrycarpus imbricatus (Blume) de Laub. **var. robustus** de Laub. *J. Arn. Arbor.* **50**: 323 (1969). Gibbs (1917: 80); de Laub.(1988: 376-380). **Type**: *L.J. Brass* 39568. 2650 m. Mt Wilhelm. Simbu Prov. Papua New Guinea.

Podocarpus imbricatus Blume
Podocarpus papuanus Ridley
Podocarpus cupressina

Ecol.: Common in rain forest from (500-) 700-3000 m. Particularly in lower montane *Lithocarpus-Castanopsis* forest and in mid montane forest dominated by *Nothofagus* and *Phyllocladus*. A canopy tree, sometimes emergent.
Dist.: Widespread from Philippines and Borneo to New Guinea.
Specimens: NETTOTI RANGE: *Versteegh* BW 10411; *van Royen & Sleumer* 7948; *Davis* 804. KEBAR VALLEY: *van Royen (& Sleumer)* 3895. ANGGI LAKES: *Gibbs* 5540; *Versteegh* BW 250; *Gjellerup* 1148 (B); *Kostermans* 2197 (L, BO); *Stefels* BW 2014, BW 2006; Lake Gigi - *Kanehira & Hatusima* 13933; Lake Gita - *Kanehira & Hatusima* 13446, 13735. ARFAK MTS: Beccari (1875); Stefels BW 2038; *Kanehira & Hatusima* 13933; Versteegh BW 12610; *Kostermans* 2197, 2269. RANSIKI: *Mangold* BW 2263. WITHOUT LOCALITY: *Beccari* 9842; BW 10411.
Notes: de Laubenfels (1988: 377) recognised 4 varieties in this species. Only var. *robustus* occurs in the Manokwari Kabupaten.

Dacrycarpus steupii (Wasscher) de Laub. *J. Arn. Arbor.* **50**: 328 (1969). de Laub. (1988: 381).

Podocarpus steupii Wasscher *Blumea* **4**: 405 (1941). **Type**: *NIFS* bb 22857. Celebes.,

Ecol.: Locally common in disturbed rain forest and poorly drained forest from (860-) 1500-2000 (-3420) m altitude.
Dist.: Borneo to Sulawesi; throughout New Guinea.

Specimens: AIFAT VALLEY: *Moll* BW 12820. WITHOUT LOCALITY: BW 12820; BW 12840; BW 12876.

DACRYDIUM Solander ex Forst. f.

Dacrydium beccari Parl. in DC. *Prod.* **16**(2): 494 (1868). de Laub (1988: 366); van Royen (1979: 34). **Lectotype**: *Beccari* 2385. 1500 m. Mt Poe, Sarawak.

Ecol.: Common in forest on mossy ridges, often dominant. From 600 to 2500 m altitude.

Dist.: Widespread in Malesia, to Solomon Islands.

Specimens: AIFAT VALLEY: *Moll* BW 12853; *van Royen & Schram* 7791.

Dacrydium cornwalliana de Laub. *Flora Malesiana* Ser. 1. Vol. **10**: 366 (1988). **Type**: *Versteegh* BW 3041. Wissel Lakes, Irian Jaya.

Dacrydium nidulum de Laub. var. *araucarioides* de Laub. *J. Arn. Arbor.* **50**: 295. f.36 (1969). **Type**: *Versteegh* BW 3041. Wissel Lakes, Irian Jaya. (L).

Ecol.: Dominant in nearly pure stands in swamp forest (Johns 1982) and possibly heath forest. 1450-2300(-2770) m.

Dist.: West and Central New Guinea.

Notes: de Laubenfels (1988) maps this species from the Vogelkop without reference to any collection.

Dacrydium nidulum de Laub. var. ***nidulum*** *J. Arn. Arbor.* **50**: 292. f.3a (1969). de Laub. (1988: 365). **Type**: *Vink* BW 15271. Vogelkop, Irian Jaya. (L).

Ecol.: Canopy tree in rain forest from 0-1200 m. Most collections below 600 m altitude from wet areas.

Dist.: Sulawesi and Moluccas, throughout New Guinea. Common in Irian Jay - elsewhere populations are often isolated.

Specimens: SEGIOR: *Vink* BW 15271; *Versteegh* BW 7378. AJAMARU: *Versteegh* BW 4985; *Vink* BW 15249. KEBAR VALLEY: *Versteeg & Kalkman* BW 5594; *Sidje* BW 5596; *Koster* BW 6885.

Dacrydium novo-guineense Gibbs *Contr.Phytogeog. & Flora Arfak Mts.* 78, f.3 (1917). de Laub. (1969: 286); de Laub. (1988: 364). **Type**: *Gibbs* 5648. 2700 m. Arfak Mts. Irian Jaya. (Holotype: BM).

Ecol.: From (700-)1500-2200(-3000) m altitude. In montane forest. At higher elevations a common small tree.

Dist.: Sulawesi and Moluccas to West and Central New Guinea.

Specimens: TAMRAU MTS: *van Royen & Sleumer* 7219. KEBAR VALLEY: *van Royen* 3857. MT NETTOTI: *van Royen & Sleumer* 7948. ARFAK MTS: *Gibbs* 5648 (Type); *Gibbs* 5508; *Gjellerup* 1032. ANGGI LAKES: *Versteegh* BW 256, BW 262, BW 269; Stefels BW 2015, BW 2033; *van Royen & Sleumer* 3857; Lake Gita - *Kanehira & Hatusima* 13501, 13518, 13987, 14106. KOEBRE MT.: *Johns* 8027; *Johns* 8047.

Dacrydium xanthandrum Pilger *Bot. Jahrb.* **69**: 252 (1938). de Laub. (1969: 304); de Laub. (1988: 369). **Lectotype**: Clemens 4504. 1800 m. Ogeramnang, Morobe Dist. New Guinea.

Ecol.: Locally common to dominant on mossy ridges from (500-)100-2700 m altitude.

Dist.: Widespread Malesian species.

Specimens:

FALCATIFOLIUM de Laub.

Falcatifolium papuanum de Laub. *J. Arn. Arbor.* **50**: 312, f.6. (1969). de Laub. (1969: 274); de Laub. (1988: 373). **Type**: de Laubenfels P483. 1950 m. Mt Kaindi, Edie Ck. Morobe Province. Papua New Guinea.

Ecol.: An understorey tree in mid montane forest with *Nothofagus* and various podocarps.

Dist.: New Guinea.

Specimens: MT NETTOTI: *van Royen & Sleumer* 8203a. ANGGI LAKES: BW 250; BW 2006; BW 2014; BW 2038. RANSIKI: BW 2263. WITHOUT LOCALITY: *van Royen & Sleumer* 7948.

Notes: The manuscript name, *Falcatifolium sleumeri,* was given to *van Royen and Sleumer* 8203.

NAGEIA Gaertn.

Leaves of *Nageia* can be confused with *Agathis*. In *Nageia* the terminal bud is long and narrow-pointed. This contrasts with *Agathis* where the terminal buds are short and distinctly rounded.

Nageia wallichiana (Presl) O.K. *Dev. Gen. Pl.* **2**: 800 (1891). Wasscher (1941: 417); de Laub. (1988:).**Type**: Blume s.n. Mt Salak, Java. (L. BO).

Podocarpus wallichianus Presl
Decussocarpus wallichianus (Presl) de Laub.
Podocarpus blumei Endl.
Podocarpus latifolia Blume *non* Thunberg

Ecol.: Common tree in lowland rain forest rarely to 2100 m altitude. Common on ridges with *Anisoptera*, *Araucaria*, *Podocarpus* and various Fagaceae.

Dist.: Widespread from SE Asia to New Guinea.

Specimens: WARSAMSON VALLEY: *Moll* BW 11652. KEBAR VALLEY: *Schram* BW 7951; *Smit* BW 2314; Sijde BW 2314, BW 5579; *Mangold* BW 2350;*van Royen (& Sleumer)* 5058. SIDAI: *Schram* BW 1785, BW 6174. MANOKWARI: *Menusefer* BW 8180. RAMOI: *Beccari* s.n. ANDAI:*Beccari* s.n. WITHOUT LOCALITY: *Beccari* 365; BW 7948.

PODOCARPUS L.'Herit. ex Persoon

Foxworthy, F.W. (1911). Philippine Gymnosperms. *Philipp. J. Sci. Bot.* **6**(3): 149-178. Pl.26-33.

Gray, N.E. (1958). A taxonomic revision of *Podocarpus* XI. The South Pacific species of section *Podocarpus*. *J. Arn. Arbor.* **39**: 424-477.

Laubenfels, D.J. de (1985). A taxonomic revision of the genus *Podocarpus*. *Blumea* **30**(2): 251-278.

Wasscher, J. (1941). The genus *Podocarpus* in the Netherlands Indies. *Blumea* **4**: 359-481.

Podocarpus archboldii N.E. Gray *J. Arn. Arb.* **39**: 452 (1958). de Laub. (1980: 140); de Laub. (1985: 266); de Laub. (1988: 402). **Type**: *L.J. Brass* 13121. Idenburg River. Berhard camp. Irian Jaya. (L)

Ecol.: Scattered to locally common in mixed lowland rain forest with *Castanopsis*. 720-2200 m altitude.

Dist.: Scattered throughout New Guinea.

Specimens: KEBAR VALLEY: *Koster* BW 7103. Tobi Mts. - *Sijde* BW 5580.

Podocarpus glaucus Foxw. *Philip. J. Sci. Bot.* **2**: 258 (1907). de Laub. (1985: 272); de Laub. (1988: 411). **Type**: Merrill 5672. 11/1906. 2400 m. Mt Nalcon, Mindoro. Philippines.
Ecol.: Locally common along ridges, 1100-2800 m altitude.
Dist.: Philippines and Sulawesi to New Guinea and Solomon Islands.
Specimens: ARFAK MTS.: *Kostermans* 2161; 2236. WITHOUT LOCALITY: BW 10382; *van Royen & Sleumer* 8203.
Notes: cf. **P. glaucus** *van Royen* 7171.

Podocarpus ledermannii Pilger *Bot. Jahrb.* **54**: 210 (1916). Gray (1958: 447); Wasscher (1941: 456); de Laub (1980: 140); de Laub. (1985: 268); de Laub. (1988: 405). **Syntypes**: Ledermann 9878, 9943, 9996, 10064. Augusta Fluss, Lordberg. 1000 m. Papua New Guinea. (B).
Podocarpus idenburgensis N.E. Gray, *J. Arn. Arbor.* **39**: 447 (1958).
Podocarpus ridleyi N.E. Gray, *J. Arn. Arbor.* **39**: 46 (1958).
Ecol.: Locally common in rain forest to 1800 m altitude.
Dist.: New Guinea, Java and Borneo.
Specimens: TAMRAU MTS.: *van Royen, Sleumer & Schram* 7792. MANOKWARI: Mt Krabo - *Koster* BW 10753. WITHOUT LOCALITY: BW 10753.

Podocarpus levis de Laub. *Blumea* **24**: 496 (1978). de Laub. (1978: 496); de Laub. (1985: 267); de Laub. (1988: 403). **Type**: *van Dijk* bb 30484. Japen Is. Irian Jaya.
Ecol.: Scattered to locally common in rain forest from 0-1650 m altitude.
Dist.: E Borneo to Irian Jaya.
Specimens: ANGGI LAKES: BW 282.

Podocarpus neriifolius D.Don ex Lamb. *Gen. Pinus* ed. **1**, 2: 21 (1824). de Laub. (1985: 266); de Laub. (1988: 400).
Ecol.: Along ridges, sometimes on ultrabasics. With *Anisoptera* and *Flindersia*.
Dist.: Widespread Asian species to Solomon Islands and Fiji. Locally common in New Guinea.
Specimens: ARFAK MTS.: *Kostermans* 2410.
Notes: Foxworthy (1941: 437-456) published a detailed study of this species within which he recognised several varieties. None of the varieties were recognised from the Kepala Burung.

Podocarpus pilgeri Foxw. *Philip. J. Sci.* Bot. 2: 259 (1907). de Laub. (1985: 272); de Laub. (1988: 410).
Ecol.: Often in *Nothofagus* forest as an understorey tree.
Dist.: S China to New Guinea.
Specimens: ANGGI GITA: *Kostermans* 2161; *Kostermans* 2236; *Kostermans* 2519. WITHOUT LOCALITY: *van Royen & Sleumer* 7958.

Podocarpus polystachys R.Br. ex Endl. *Syn. Conif.* 215 (1847). de Laub. (1985: 276); de Laub. (1988: 418). **Type**:
Podocarpus thevetiifolia
Ecol.: From sandy beaches to degraded heath forest and in New Guinea on limestone (Kepala Burung).
Dist.: Scattered throughout Malesia to west Irian Jaya.
Specimens: WITHOUT LOCALITY: *Beccari* 9852.

Podocarpus pseudobracteatus de Laub. *Blumea* **26**: 142 (1980). de Laub. (1985: 271); de Laub. (1988: 409). **Type**: *Kalkman* 5189. Mt Ambua. Souhtern Highlands. Papua New Guinea.

Podocarpus archboldii var. *crassiramosus* N.E. Gray *J. Arn. Arbor.* 453 (1958). **Type**: *L.J. Brass* 10874. Lake Habbema. Irian Jaya.

Ecol.: Scattered in lower montane and mid montane forest dominated by *Castanopsis* and *Lithocarpus*. Also in *Dacrydium* swamp forest. 1740-2850 m.
Dist.: New Guinea.
Specimens: ARFAK MTS. : Putat - *Gibbs* 5985.

Podocarpus rumphii Blume *Rumphia* **3**: 214 (1847). de Laub. (1985: 275); de Laub. (1988: 415).
Ecol.: Locally common in rain forest from 5-200(-1550) m.
Dist.: Widespread Malesian species.
Specimens: ARFAK MT.: *Beccari* 9848. KAPAOR: *Beccari* 78.

Podocarpus rubens de Laub. *Blumea* **30**(2): 266 (1985). de Laub. (1988: 401). **Type**: *Womersley & Millar* NGF 6980. Waghi-Sepik Divide. Papua New Guinea.

Podocarpus neriifolius var. *timorensis* Wasscher (1941: 451).
Podocarpus pilgeri van Royen *non sensu* Foxw. (1979: 29).

Ecol.: Medium sized subcanopy tree mostly above 1500 m altitude. Often in fagaceous forests.
Dist.: Widespread Malesian species.
Specimens: WITHOUT LOCALITY: BW 11456.

PRUMNOPITYS Philippi

Prumnopitys amara (Blume) de Laub. *Blumea* **24**: 190 (1978). Wasscher (1941: 381); de Laub. (1988: 385).

Podocarpus amarus Blume *En. Pl. Java.* **1**: 88 (1827).

Ecol.: Scattered to common in rain forest. Often common in fagaceous forest. (0-)500-2000(-2300) m altitude.
Dist.: Throughout Malesia to coastal Queensland.
Specimens: RANSIKI: BW 2245. WITHOUT LOCALITY: BW 2318; BW 2355; BW 5578; BW 7785; BW 9527; BW 10354; BW 10402; BW 11336; BW 12826; *van Royen & Sleumer* 6719.

GNETACEAE

Markgraf, F. (1930). Monographie der gattung *Gnetum*. *Bull. Jard. Bot. Btzg.* III **10**: 407-511.

Markgraf, F. (1951). Gnetaceae. *Flora Malesiana Ser. 1.* **4**: 336-347.

Markgraf, F. (1972). Addenda, corrigenda et emendanda. Gnetaceae. *Flora Malesiana Ser. 1.* **10**: 944.

GNETUM L.

Gnetum gnemon L. *Mant.* **1**: 125 (1767). Markg. (1930: 436); Markg. (1951: 340).

var. gnemon

Dist.: Throughout Malesia to Solomon Islands and Fiji.
Ecol.: Understorey tree in forest to 10 m tall. Sea-level to 1000 m. Not rare. Often planted as a food (seed and leaves) and fibre source.

Specimens: ARFAK PLAINS: *Sands* 6300. ARFAK MTS.: Mupi Valley - *Davis* 659; Mahyar 397. ANDAI: *Beccari* 525; *Beccari* 9783. WAREN: *Kanehira & Hatusima* 14237. WITHOUT LOCALITY: *Beccari s.n.*; BW 11333;

Gnetum gnemonoides Brongn. in Duperrey *Voy. Coquille* 12 (1829). Markg. (1930: 480); Markg. (1951: 341). **Type**: Rumphius, Herb. Amboinense 5: taf. 7 (1747).
Gnetum rumphianum Beccari *Malesia* **1**: 182 (1877).
Ecol.: In rain forest at low altitudes, to 300 m.
Dist.: Widespread throughout Malesia to New Guinea.
Specimens: RAMOI: *Beccari* s.n.; *Beccari* 9779. ANDAI: *Beccari* 526; *Beccari* 9782. MANOKWARI: Gunung Meja - *Widjaja* & 3126. WITHOUT LOCALITY: *Beccari* s.n.; BW 11346.

Gnetum latifolium Blume *Tijd. Nat. Geschied & Phys.* **1**: 160 (1834). Markg. (1930: 458); Markg. (1951: 342).
Ecol.: Locally common climber to 1800 m altitude.
Dist.: Widespread throughout Malesia.
Specimens: ARFAK MTS.: *Beccari* 9776. ANDAI: *Beccari* 749 (Herb. Flor. *Beccari* 9777). RAMOI: *Beccari* 363; *Beccari* 9778. WASIOR: *Widjaja* 4607. WITHOUT LOCALITY: *van Royen & Sleumer* 6721; BW 7736; BW 11398; *Beccari* s.n.
Notes: Markgraf (1930: 458-466) includes a detailed discussion of several varieties and forma which he originallyt recognised in this species.

***Gnetum* sp. 1**
Dist.: Known from only a single collection in Kepala Burung.
Ecol.: Extensive woody climber in partially disturbed lowland rain forest.
WARIORI PLAIN: Inland Masni Camp - *Sands* 6183.

***Gnetum* sp. 2**
ARFAK MTS.: Mupi Valley System - U.W.Mayar 397.

***Gnetum* sp. 3**
WITHOUT LOCALITY: BW 12857.

***Gnetum* sp. 4**
WITHOUT LOCALITY: BW 12584.

Gnetum spp. KEBAR VALLEY: *Johns* 8292. RAMOI: *Beccari* 325 (FL). ARFAK MTS.: Putat - *Beccari* 924 (FL). NUNI: Sungai Asai - *Dransfield, J. & Zona* 7592.

CPSIA information can be obtained
at www.ICGtesting.com
Printed in the USA
BVOW07s1225140616
451621BV00010B/11/P